Epic Sea Battles

Great Battles for Boys

Joe Giorello
with
Sibella Giorello

Epic Sea Battles

Print Edition

This book is dedicated to two Navy seamen: my father-in-law, Alaska Supreme Court Justice Roger G. Connor, veteran of World War II; and my brother-in-law, John Quinn, Gunnersmate 1967–71.

Thank you for your service.

TABLE OF CONTENTS

INTRODUCTION

FOR THOUSANDS OF YEARS, the seas have been more than just highways for trade and exploration—they've been battlefields that shaped the course of history.

From ancient times to the modern era, control of the oceans has often meant the difference between victory and defeat.

In this book, you'll discover how naval warfare began with wooden ships powered by oars and sails, battles in which ramming and hand-to-hand combat were key tactics. As you turn the pages, you'll see how technology transformed the seas into even deadlier arenas—bringing in steel warships, powerful cannons, aircraft carriers, and submarines.

Sea battles are worth studying because they show us how human ingenuity, courage, and strategy have clashed on a global scale. These battles have determined the fate of empires, protected nations, and changed the way wars are fought.

While there are far too many historic sea battles to cover in a single book, I've chosen a mix of famous and lesser-known clashes to give you a broad view of naval warfare through the ages. You'll notice that the last truly massive sea battles happened during World War II—after that, high-tech weapons like guided missiles allowed ships to strike from far away, changing the face of naval combat forever.

So get ready to dive in.

The ocean is calling, and its history is filled with epic stories.

P.S. Two items:

First, at the back of this book, I've provided a glossary with definitions of naval terms. Make sure you check it as you're reading.

Second, it might sound weird to call a ship "she," but this tradition goes back thousands of years. Sailors often considered their ships protective and nurturing, like a mother. Also, many languages (Latin and French, for example) refer to objects as "she" and "he."

In English, that doesn't happen often—except for ships. So don't be surprised when a ship is referred to as "she."

Now, let's get into some naval battles!

The Battle of Mylae

Second Punic War

260 BC

Artist's depiction of Rome and Carthage in naval warfare during the Punic Wars. Note the archers (left) and legionaries with shields and swords (right). Also, the unfortunate soldier falling overboard (center).

HAVE YOU HEARD OF the Punic Wars?

There were three Punic Wars during the years 264 BC to 146 BC. These three wars were fought between Ancient Rome and Carthage, located across the Mediterranean Sea in today's North Africa.

At the time, these Punic Wars were the largest and most deadly battles on earth. Hundreds of thousands of soldiers died on both sides.

What was all this fighting for?

Both Rome and Carthage wanted to rule the Mediterranean Sea and the lands surrounding it.

Here's a basic map of the area at that time. The Mediterranean Sea separates Rome (modern-day Italy) from Carthage (modern-day North Africa). Notice that off the "boot" of Italy is the triangle-shaped island of Sicily.

The First Punic War began in 264 BC, with the Romans attacking and gaining ground on parts of Sicily. Ambitious as ever, the Romans wanted to take the entire island—Sicily was a key location for trade and military dominance.

Their problem was Sicily is an island, and the Romans didn't have a navy.

The Carthaginians had a powerful navy.

How can you defeat an enemy on water in a naval battle if you don't have an actual navy?

To solve their problem, the Romans captured a Carthaginian ship, called a quinquereme, and copied its design.

Quinqueremes were massive wooden ships with five levels of rowers. (In Latin, "quinque" means "five.") From bow to stern—front to back—the ships stretched about 150 feet in length.

To grasp the size of these ships, walk forward 150 steps. Now imagine several levels of rowers on either side, pulling in unison to speed that ship over the water.

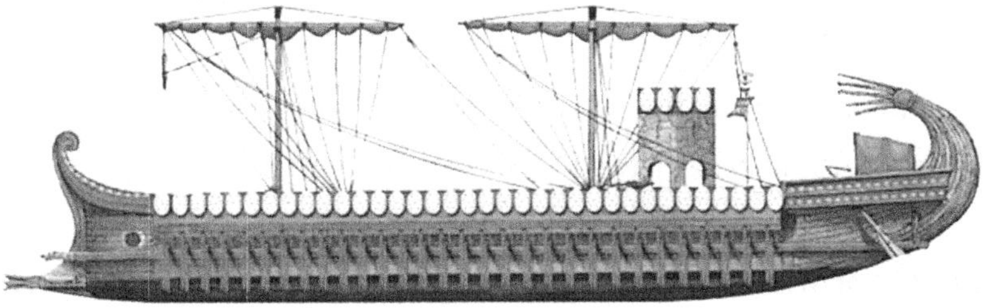

Example of a quinquereme.

But the Romans didn't just copy that Carthaginian ship. They invented something original to go with it.

On land, the Roman army often defeated the Carthaginians, so they needed to apply those land-based strategies and tactics to the sea.

The Romans were excellent military innovators, a trait that helped their legions conquer most of the known world.

The Roman navy's first defining clash was the Battle of Mylae.

The Carthaginian naval strategy involved charging in formation, then plowing through enemy weak points and ramming into ships to create holes that let in water and sank the vessels.

The strategy was very effective!

The Carthaginians would also aim for an enemy ship's oars to hinder their ability to maneuver—or escape.

In addition, they launched flaming arrows and hurled fire-filled clay pots onto the enemy's wooden decks. Then, when they got close, the Carthaginians used grappling hooks to pull the ships alongside and climb aboard for brutal hand-to-hand combat.

The Carthaginian crews were so highly trained in naval warfare that they could ram ships with just enough force to puncture the enemy but not become stuck themselves. That was quite a challenging task since the ship was bobbing up and down on the churning waters of the battlefield. That kind of skill and precision required years of training.

At Mylae, the Carthaginian navy was commanded by a man named Hannibal Gisco—not the more famous Hannibal who would later invade Italy with elephants during the Second Punic War (read about the other Hannibal in my *Ancients to Middle Ages* book).

Hannibal Gisco.

Hannibal Gisco had already won a recent naval battle against Rome. That loss convinced Roman general Duilius that his crews could not win this battle using conventional naval tactics.

Duilius's only path to victory was to transform this sea battle into a land battle.

How would he do that?

With a new weapon: the corvus.

The corvus was a mounted boarding ramp or drawbridge—about 35 feet long and 4 feet wide, much like a gangplank. But it could also swivel from side to side and featured an iron hook at its far end that resembled a bird's beak. In fact, that's how the corvus got its name. In Latin, "corvus" means "crow" or "raven."

Artist's rendering of what the corvus might have looked like.

With that hook on its end, the corvus could latch onto an enemy ship, allowing Roman soldiers to rush across the four-foot-wide plank and engage in hand-to-hand combat—a domain where the Roman legion excelled.

By the time of the Battle of Mylae, the Romans had built up a fleet of about 100 ships, mostly quinqueremes, along with some triremes (three-oared vessels).

The Carthaginian navy had about 130 ships.

Now imagine the Mediterranean Sea around the island of Sicily, choked with hundreds of warships, their oars thrashing the water as they sped toward each other. On deck, armed sailors and soldiers stood ready to inflict deadly harm on their avowed enemy.

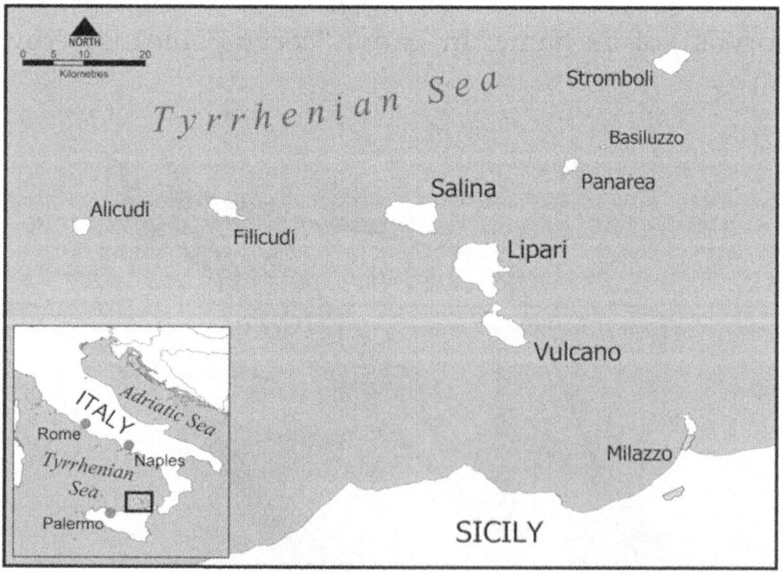

Area of the battle off Sicily, located near the "toe" of Italy's "boot."

Confident in their skills and training, the Carthaginians advanced. Naval warfare was their arena. The sailors watched the Roman fleet sail closer and closer and then …

BAM!

The Carthaginian ships rammed the Roman vessels.

They hit from the side! From the rear!

Oars snapped!

Hulls cracked and flooded with seawater!

But just as quicky, the Romans deployed their new weapon, the corvus.

The swiveling ramps locked the enemy ships in place. Roman legionaries stormed across the bridge and leapt onto the Carthaginian decks, transforming a sea battle into a deadly conflict. Swords slashed, knives stabbed, shields shoved. Men screamed and fell overboard, drowning in the raging sea.

Shock and terror swept through the Carthaginian crews. They were completely unprepared for this new tactic.

Roman legionaries continued to charge forward, hollering war cries just as they did on land. The corvus soon helped the

Romans seize thirty Carthaginian ships, including the flagship carrying Hannibal, who managed to escape.

The Carthaginian ships attempted to retreat, or even flank the Romans. But the corvus could pivot and latch onto attackers from nearly every angle.

Twenty more Carthaginian ships were hooked and captured.

Hannibal fled the battlefield and raced away with the remnants of his fleet. The Battle of Mylae was over.

Rome had won its first-ever naval victory!

The Romans continued to pursue the fleeing Carthaginians all the way to Sardinia, where they destroyed even more enemy ships. The defeat was so devastating that Hannibal was later arrested by his own men in Carthage and crucified for his failures.

Illustration of Hannibal arrested by his own men.

The First Punic War continued after the Battle of Mylae, lasting a total of 24 years. According to Polybius, a Greek historian who wrote about Rome, this war cost Carthage 500 ships. Rome lost 700.

Eventually, Rome won the First Punic War. The victory granted it critical territory that helped establish the empire as the most powerful force in the known world. And, after this battle, the Roman navy was recognized as much as its fierce legion.

Interestingly, while the corvus gave Rome a critical edge early in the war, it was later abandoned due to the top-heavy instability which didn't work that well in stormy seas.

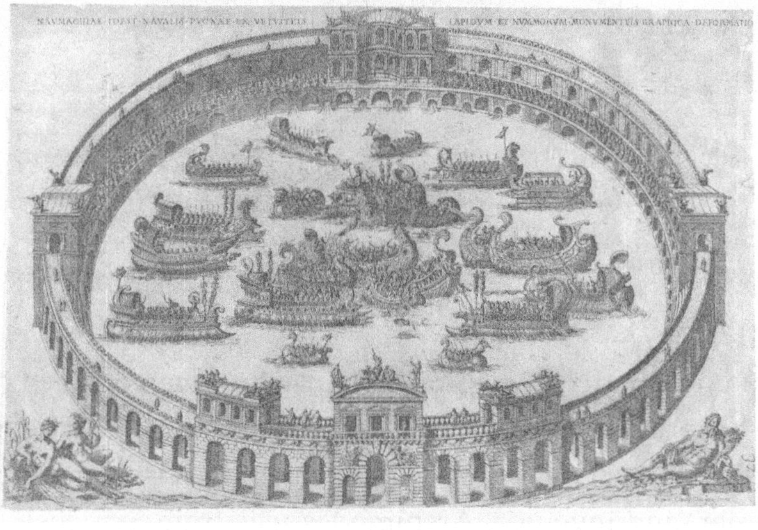

The Roman Colosseum with staged naval battles.

But Rome became so proud of its navy that later emperors staged naval battles right in the middle of the city!

Using brilliant engineering methods, the Romans diverted water from the nearby Tiber River to flood the famous Colosseum—normally used for gladiator battles. Entire sea battles were recreated there for the Colosseum audiences to "witness" the clashes.

WHO FOUGHT?

Archimedes Thoughtful (also known as *Portrait of a Scholar*), a painting by Domenico Fetti, 1620.

Although nobody knows for certain who invented the corvus, some historians suspect it was Archimedes ("ark-uh-me-dees").

An Ancient Greek mathematician, physicist, engineer, astronomer, and inventor, Archimedes was from Syracuse, a city on the island of Sicily.

Many details of his personal life are unknown, but Archimedes is considered the greatest mathematician of all time. His pioneering contributions to math can still be seen in today's modern calculus, geometry, and measurements of objects such as spheres.

During his lifetime, however, Archimedes was most famous for designing innovative machines, including defensive war mechanisms that helped protect his native city of Syracuse from invasion.

Polybius, the ancient historian, wrote about Archimedes's machines, including how he improved catapults and designed crane-like mechanisms that could swing around in an arc.

Though Polybius doesn't mention the corvus specifically, the brilliant design seems like something that only Archimedes could imagine.

After his death, Archimedes was placed in a tomb adorned with an inscription that showed his favorite mathematical proof—that a sphere's volume and surface area are two-thirds the size of an enclosing cylinder, including its bases.

Remember that (and Archimedes) when you study geometry!

BOOKS

Tools of the Ancient Romans: A Kid's Guide to the History & Science of Life in Ancient Rome by Rachel Dickinson. Apart from an overview of Roman life, this book also includes hands-on activities, such as making a mosaic and an abacus.

INTERNET

Roman boats and transportation: https://www.historyforkids.net/ancient-rome/daily-life/roman-boats-and-transportation.html/

More on the triremes, which resemble quinqueremes: https://online.kidsdiscover.com/quickread/triremes-triple-decker-warships-that-ruled-the-ancient-seas

MOVIES

Roman Feats of Engineering (2007), available on Prime Video. The marvels of Ancient Rome's practical engineering, roads, arches, and aqueducts are presented in this program. Reenactments, diagrams, and live-action footage examine the awe-inspiring accomplishments.

Superweapons of the Ancient World (2005). Archimedes is believed to have designed a defensive machine known as "the ship shaker"—a claw consisting of a crane-like arm with a large metal grappling hook, much like the corvus. But when the claw was dropped onto an attacking ship, the arm swung upward, lifting the ship out of the water in order to sink it. The creators of this television documentary built a version of the claw and found it was a workable device. Available on YouTube: https://youtu.be/GHkXIw_n7CE?feature=shared

The Siege of Paris

AD 885–886

Illustration of a Viking warrior, landing his longboat for a raiding party.

THE WORD "VIKING" IN OLD NORSE language meant "pirate" or "raider."

Today, we remember them as fierce, seafaring warriors. But they were also explorers, settlers, and traders. The Vikings hailed from Scandinavia, which is modern-day Norway, Sweden, and Denmark.

Their most famous vessels were longships. These sleek, fast boats were equipped with both sails and oars, built from sturdy oak, and designed with shallow hulls (bottoms), which made them ideal for river travel as well as open sea voyages.

The Vikings' larger longships could carry about 100 warriors, while smaller versions of those vessels held a dozen or so men, ideal for scouting missions and quick raids.

The longship's most noticeable feature was the dragon head carved into the prow (the front of the boat). These terrifying decorations were meant to scare enemies—and perhaps even sea monsters.

Illustration of a typical Viking longboat.

The Vikings traveled far and wide from their far northern home. How did they manage to navigate to far distances on the ocean without modern tools such as compasses or GPS?

The Vikings studied the stars and followed the constellations, as ancient mariners have done for centuries. The Vikings could also read wave patterns, track the behavior of tides, and notice how the color of the sea changed with depth. They also followed birds—especially migratory ones that signaled land was nearby.

But before they were fearsome warriors of the north seas, most Vikings were just farmers and fishermen. However, Scandinavia's rocky, cold lands weren't great for growing crops. As their population grew, many young men took to the seas in search of food, wealth, and adventure.

During their travels, Vikings discovered wealthy monasteries along the coasts of Europe—places where Christian monks lived, worked, and prayed. These monasteries contained gold, silver, and food, but very few soldiers.

To the Vikings, the monasteries were easy targets. So were other parts of Europe.

One of the most daring and dramatic Viking attacks was the Siege of Paris in AD 885.

Why did the Vikings attack Paris?

Since around 841, Vikings had been raiding towns and villages along the Seine River, which was a perfect highway for their longships.

The Seine also flows through Paris. By 885, after raiding up and down the river, the Vikings wanted more than just loot—they wanted to take Paris itself.

So they gathered a massive invasion force: 700 longships, carrying anywhere from 10,000 to 40,000 Viking warriors.

At the time, the man ruling that part of France was King Charles the Bald (yes, that was really his name!).

Charles the Bald, as depicted sometime around 845.

Instead of fighting the Vikings, Charles had been paying them something called Danegeld, or protection money, to keep the Vikings from ravaging his land.

Charles the Bald's strategy totally backfired.

All those payments only encouraged the Vikings to come back for more.

And now they wanted Paris, too.

On the other hand, you can't really blame baldy Charles for being afraid: the Vikings were terrifying.

They were strong, muscular, and unusually tall for their time—often over six feet. Historians believe their height came from their protein-rich diets of meat and fish. They had long hair and beards, fierce blue or gray eyes, and wore wool tunics, pants, leather belts, and battle scars earned through years of raiding.

They were well armed, too. Here were their favorite weapons:

- Battle axe – also known as the Dane axe. Razor sharp, this weapon could cut through helmets and shields.
- Ulfberht sword – a high-quality, double-edged steel sword about three feet long. The best swords carried this mysterious name.
- Spears – for throwing or stabbing in combat.
- Seax – a short sword used for close-quarter fighting.
- Shield – round, made of wood with iron reinforcements. It was about three feet across and very heavy.
- Helmet – iron-made. Despite what you see in cartoons and movies, real Viking helmets didn't have horns.

Artist's rendering of what a Viking probably looked like.

Picture it: here comes a fleet of dragon-headed longships packed with strong, frightening warriors, carrying deadly weapons, sailing straight toward you—because *you* were their target! It must've been a terrifying sight.

However, the city of Paris also had a hero: Count Odo.

While Charles the Bald was king of the greater area, Odo was the local ruler of Paris. And unlike the king, Odo refused to pay off the Vikings.

At the time, Paris was a fortified city on an island in the Seine River. Strong walls made of wood and stone surrounded the city. Two fortified wooden bridges connected the island to the riverbanks on either side, each bridge protected by stone towers. This location gave Paris a natural moat, making it a tough target for any army. (A moat is a deep, wide ditch, usually filled with water, and typically surrounding fortified medieval towns, fortresses, and castles to protect against assault.)

When the Vikings sailed the Seine to Paris, they were led by two powerful warlords: Siegfried and Rollo.

Their first target was the northern bridge, which they hoped to break down so they could storm the city.

Imagine wave after wave of these thousands of Viking warriors charging the bridge, climbing up ladders, hurling massive stones and flaming tar from catapults, and smashing at gates with battering rams.

Example of a medieval catapult.

The Parisians fought back from their positions on the walls, pushing the ladders down, firing arrows, and pouring boiling oil and hot wax on their attackers.

Many Vikings fell into the river below, where their ships were waiting.

The Vikings brought more siege engines—large war machines that were like towers on wheels—and battering rams used to break through walls.

But nothing worked.

So the Vikings changed tactics.

They surrounded the city and laid siege. That means no one could get in or out.

No supplies.

No food.

No reinforcements.

Nothing.

The Vikings built more siege towers and even tried to tunnel under the city walls to make them collapse.

But the Parisians refused to give up.

Count Odo led night raids on the Viking camps. He and his men snuck out of the city and struck the Viking under cover of darkness, then retreated before dawn.

The fighting went on for months, all the way into AD 886, when the Vikings finally destroyed the northern bridge. It crashed into the river with a roar.

But the city still didn't fall.

Exhausted and frustrated, the Vikings were growing desperate.

Then something strange happened. A new king, Charles the Fat (yes, really), didn't send troops to rescue Paris. Instead, he did what Charles the Bald did—he paid the Vikings.

He offered them a huge amount of silver and begged them to leave.

And they did.

After almost a year of battle, the Vikings never took Paris.

Count Odo was hailed as a hero. The people loved him for standing firm. A few years later, in 888, he became king of West Francia (which was the western part of what would one day become France).

WHO FOUGHT?

Charles the Bald was born in the year 823. He became king of part of Europe, and even though his nickname was "the Bald," we don't know if he was actually bald! It might have just been a funny name.

But here's an interesting fact: Charles the Bald was the grandson of Charlemagne, one of the most powerful rulers—and warriors—in European history.

Who was Charlemagne?

Charlemagne (which means "Charles the Great") was a mighty king who ruled over a huge empire in Europe more than 1,200 years ago. He was king of the Franks and later crowned "Emperor of the Romans" in the year 800 by the Pope.

That made Charlemagne the most important Christian ruler in Europe. He helped spread learning, built schools, and brought peace after years of chaos and warfare.

People sometimes call Charlemagne the "Father of Europe."

Charlemagne had many children, and when he died, his empire was passed down to his son Louis, who was Charles the Bald's father.

When Louis died, his sons (including Charles) fought over who would get what part of the empire. There were battles between brothers, and the land was eventually split into three parts under the terms of the Treaty of Verdun in 843.

Charles got the western part of the empire, called West Francia. That land would later grow into what we now call France.

BOOKS

The Viking Age for Kids: A Captivating Guide to the Vikings, Their Raids, and Everyday Life

Anglo-Saxons for Kids: A Captivating Guide to the People of Early Medieval England and Their Battles Against the Vikings (both by Captivating History)

The Story of Beowulf (Dover Children's Classics) by Henrietta Elizabeth Marshall

INTERNET

A good visual depiction of an earlier siege of Paris in 845 (note: this video's first 40 seconds advertise a battle game): https://youtu.be/zAInHOm0tBE?si=nzZhowlfEuep1S8X

The Battle of Dover

"The Battle of Sandwich"

August 24, 1217

THE BATTLE OF DOVER (*see page 20*).

WATERWAYS OFTEN DETERMINE A COUNTRY'S FATE, in both commerce and defense. Not only do trading ships bring in goods and supplies, but enemy nations tend to invade via waterways—using rivers and coastlines for access to the interior.

Historically, most countries focused on the waterways nearest to them. However, in this next battle, which took place in the year 1217, naval warfare would take place on the open seas, marking one of the first times in history that sailing ships fought in this way. The outcome of this battle would also determine the future of England.

The Strait of Dover separates the southeastern coast of England from the northeastern coast of France. The Strait also connects the English Channel to the North Sea.

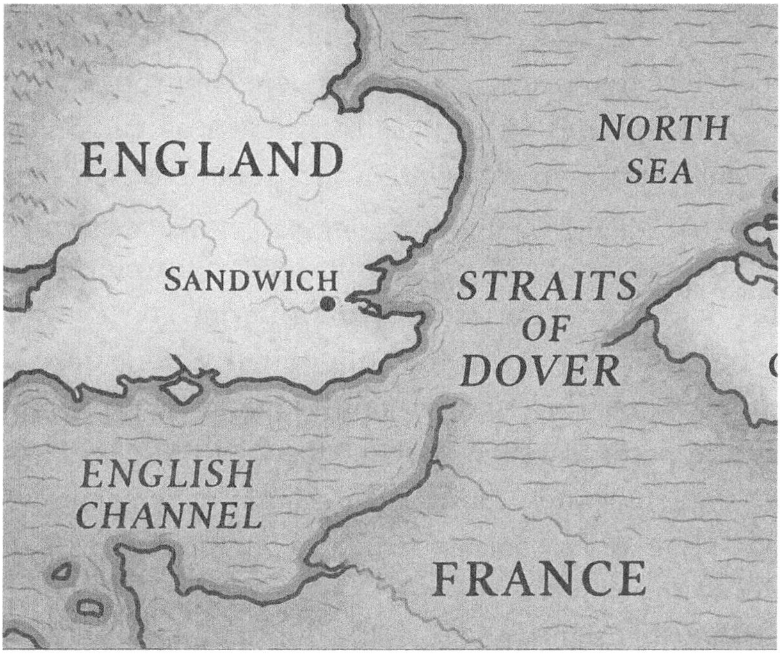

An island nation, England maintained a strong navy to boost sea trading and to provide some defense against enemy invasion, particularly from France.

The French, after all, were on a near-constant pursuit to invade English soil.

In 1217, the French cobbled together a fleet of about eighty ships for the sole purpose of invading their island neighbor to the north. This fleet was commanded by two men: Robert de Courtenay and Eustace the Monk.

Eustace got his name because he was once a monk, but he broke his religious vows to become a much-feared pirate.

Eustace also was known for switching sides between England and France—basically, whoever offered him the best deal received his "loyalty."

Meanwhile, England had a fleet of about forty ships, under the command of Hubert de Burgh, an English nobleman.

In the 1200s, most warships were actually merchant ships converted for battle. These vessels had strong, reinforced hulls designed to carry heavy cargo but were repurposed to carry sailors, archers, and siege machinery such as catapults. The ships' sails were typically large and square, while oarsmen provided extra propulsion.

At this time, England was in quite a bit of turmoil. The royal treasury had been drained by years of war, and King John's dictatorial rule had made him unpopular among his nobles. King John was a weak and unpopular monarch, known for his cruelty and pettiness. (In fact, while the tales of Robin Hood are fiction, they accurately portray King John as a villain.) Additionally, England's social structure made it difficult to overthrow any king.

It was called the feudal system, and life within it was organized like a big pyramid. At the top was the king, who theoretically owned everything—fields, forests, and villages. But he couldn't take care of all that land by himself, so he granted large pieces to lords and barons. In exchange for that land, these noblemen promised to provide soldiers to help the king in times of war.

The lords then gave smaller portions to knights, who served as warriors. The knights wore armor, rode horses, and protected the lands in times of danger.

At the very bottom of this pyramid were peasants, known as "serfs." They farmed the land, gave crops to their knight or lord, and paid taxes.

The feudal system gave the king great power, which John abused. In response to that abuse, England's barons began demanding more rights. The dispute grew so heated that in 1215, King John was forced to sign a remarkable document. It was called the Magna Carta, which is Latin for "The Great Charter."

The Magna Carta limited the king's power and established certain rights for the English nobles. (The Magna Carta would later influence documents such as the Bill of Rights in the United States, which asserts that "all men are created equal.")

Barons pressuring King John of England to sign the Magna Carta.

But just as these developments were happening, the Catholic Pope intervened.

Named Innocent III (the third), the Pope said King John didn't have to uphold his end of the Magna Carta deal.

The Pope's declaration spurred the barons to escalate their fight against King John, which became known as the First Barons' War.

The barons wanted to get rid of King John. But to do that, they'd need help.

So who did they turn to?

Their enemy—France!

The barons went to King Philip II (the second) of France. If Philip would support the barons overthrowing John, they would support him as the new king of England.

Philip didn't like the plan. He declined to help the barons.

However, his son, Prince Louis, was much more receptive to this idea. Louis quickly gathered a fleet of ships and sailed for England.

In May 1216, the French landed and began capturing key English cities and villages, including London.

Medieval image of Prince Louis of France arriving in England in 1216.

But later that year, the situation took another unexpected turn.

While on a military campaign, King John contracted dysentery, a wicked bowel infection. John died.

Now the royal throne passed to his son—Henry III.

But Henry was just nine years old!

The rise of this young king split the barons. Some switched their support back to the English monarchy, hoping to manipulate him into doing whatever they wanted. Other barons remained loyal to the idea of putting a French king into power in England.

But things don't always go according to plan, especially in war and conflict.

A knight named William Marshal, one of England's greatest warriors, became Henry's regent, or main advisor. Marshal told the young king to forget about what the Pope said and start treating the barons with more respect. While this change pleased some barons, others still wanted a French king to rule England.

And now we arrive at our next epic sea battle that would decide England's future.

Remember, by this point, the French soldiers had invaded England and taken control of several cities. But now William Marshal rallied about 500 English knights and ordered them to attack the town of Lincoln—where Prince Louis was holed up with his French soldiers.

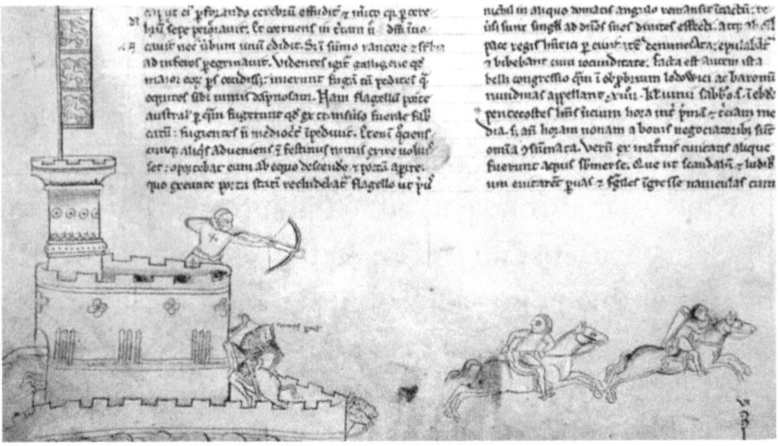

Illustration from 1217 of the Battle of Lincoln, by Benedictine monk Matthew Paris. Knights inside the castle help drive out the French forces.

Lincoln, an ancient walled city, held strategic importance militarily due to its location straddling two main roadways. And although the town was taken over by the French, the people inside Lincoln Castle remained loyal to young King Henry.

William Marshal was considered the greatest knight in England at the time. Commanding this legion of English knights, he sacked the city—with help from the loyalists inside Lincoln Castle. Marshal's knights even managed to drive Louis out of southeast England altogether.

Now on the run, Prince Louis put out a desperate call for reinforcements. That's when France assembled that mighty fleet of 80 ships—70 supply ships, 10 warships—commanded by Robert de Courtenay with Eustace the Monk at his side.

Four of the warships carried men-at-arms, basically knights, while the other six carried regular soldiers. The supply ships, which were smaller than the troop-carrier ships, were packed with horses, money for paying the French soldiers, and large pieces of war machinery, such as a trebuchet—a large catapult used to hurl boulders at enemy forces. These French ships were so overloaded that seawater splashed over the decks.

Example of a medieval trebuchet, the giant catapult that could hurl everything from boulders to fire pots at an enemy.

If this French fleet landed on English soil, the outcome made England's future questionable at best.

In defense, England assembled their fleet of 40 ships commanded by Hubert de Burgh.

French ships on the left, English on the right.

On August 24, 1217, the two fleets sailed toward each other in the Dover Strait, near the town of Sandwich—which is why this epic clash is sometimes called the Battle of Sandwich.

At first, the English fleet held back, eyeing the French ships sailing toward them with the wind at their backs. That wind was a crucial factor; it gave the French ships more speed and maneuverability in battle.

Then, suddenly, English commander de Burgh ordered his ship to attack. The flagship headed out, ready to brawl.

However, just as suddenly, de Burgh's ship veered away from the enemy.

Watching him turn tail, the French commander, Eustace the Monk, assumed the English had realized they were heavily outnumbered and lost their nerve.

Eustace pursued de Burgh.

What Eustace didn't realize was that this maneuver was a feint—a trick—where you pretend to go one way but quickly switch direction. (Professional athletes use feints all the time in football and basketball.)

The sudden change of course also shifted the wind in favor of the English fleet, giving them the edge in maneuverability. Cleverly, de Burgh then steered his ship toward the rear of the French formation and captured two French supply ships, taking the crews prisoner or killing them.

The English also deployed their archers, who fired volleys of arrows at the French sails. When a sail is torn, it can't catch wind, making the ship difficult to steer. Rip up enough of an enemy's sails and he's nearly a sitting duck. The arrows also killed many of the French crew. There were also flaming arrows that set the enemy ships on fire.

As the battle continued, the heavy warships drew so close that the crews could stare directly into the faces of their enemies.

Now the English broke out pots of powdered lime and hurled them into the wind—which was blowing straight toward the French. Highly corrosive, the lime burned their skin and blinded their eyes. Horrible screams rose, joining the chaotic sounds of sea battle: crashing water, ripping sails, splintering wood.

The battle raged on.

While the French fleet held formation, the English launched an assault on the French flagship, the *Great Ship of Bayonne*. The other ships could not offer it support.

Seizing that opportunity, the English pounced and sent more ships to surround the *Bayonne* on each side, locking her in place.

Now the English crews hurled grappling hooks and latched their vessels to the *Bayonne*.

There was no mercy.

Leaping aboard the French ships with steel swords flashing and clubs bashing, the English threw the French sailors overboard and left them to drown in the churning waters of the Dover Strait.

Witnessing this slaughter, the rest of the French fleet broke formation and fled for home.

But the English weren't done.

They turned their fury toward the French supply ships, plundering all the goods and treasure—and massacring the sailors. No prisoners were taken. That attack on the supply line, however, gave nine remaining French warships just enough time to escape.

England declared its victory.

French Prince Louis escaped and would later become King Louis VIII of France.

Eustace also got away, but he fared less well (you'll read about him below).

This epic sea battle off the coast of Dover convinced Prince Louis to abandon his mission to conquer England.

A few weeks later, he signed the Treaty of Lambeth and renounced his claim to the English throne.

WHO FOUGHT?

Eustace the Monk was born in France around 1170. In his youth, he traveled to Spain, where he reputedly studied the "dark arts," such as magic and spells.

One chronicler later wrote, "No one would believe the marvels he accomplished, nor those which happened to him many times."

Eustace eventually returned to France and entered a Benedictine monastery. The Benedictines are a group of monks who live in community and follow a set of rules established by St. Benedict.

Accounts differ on what happened next: some claim his father was murdered, prompting Eustace to leave the religious cloister to avenge him; others say his father had already died. Either way, Eustace abandoned his life as a monk and became a pirate.

Operating in the English Channel and the Strait of Dover, Eustace served both his own ambitions and acted as a mercenary—paid warrior—for England and France.

At times, King John of England employed him against King Philip II of France. Along with his brothers, Eustace led raids into northern France and even captured several islands.

But in 1212, Eustace switched allegiances and began raiding English shores. During the First Barons' War in 1215, he supported the rebel barons and ferried Prince Louis of France across the Channel to help their fight.

Though Eustace escaped the Battle of Dover, his leadership was now questionable. Some legends claim his own crew turned against Eustace and executed him by beheading. Other historians say he was captured by the English, who executed him.

BOOKS

Days of the Knights: A Tale of Castles and Battles by Christopher Maynard

The Middle Ages for Kids: Through the Lives of Kings, Heroes, and Saints by Catherine Fet and others

King Arthur: Tales from the Round Table (2020) by Andrew Lang

INTERNET

More about the Battle of Sandwich from the Guildhall Museum in Sandwich, England: https://www.youtube.com/watch?v=ox6qCXKw7_o

Want to build a medieval trebuchet? Here's how: https://www.redstoneprojects.com/trebuchetstore/trebuchet_plans.html

The Battle of Medway

June 12–14, 1667

IN THE MID-1600S, England had the most powerful navy in the world. But it was locked in a fierce battle for control of ocean trade routes with Holland, a part of the world we now know as the Netherlands.

To fight back against the Dutch (the people of Holland), England passed the Navigation Act in 1651. This law said that only English ships could bring goods into England. The law also covered English colonies—like those in America and the Caribbean—which meant they could only trade with England, not with other countries, such as Holland.

While all this fight over trade and laws was happening, England was hit by several terrible disasters. In 1665, the Bubonic Plague swept through the country. Spread by the fleas of rats, this disease killed between 60,000 and 100,000 people. And then, just two years later, in 1667, the Great Fire of London destroyed over 13,000 buildings. The fire burned for four days straight.

These disasters nearly bankrupted England. There was barely any money to pay for the Royal Navy—or even the workers who built and repaired the navy's ships.

Samuel Pepys, then Secretary of the Navy, said England was suffering from "incapacity through a total want of money."

That's a fancy way of saying, "We're totally broke."

English sailors started quitting the Royal Navy and joined merchant ships' crews instead. Dockworkers, some of whom hadn't been paid in a year, began stealing ship supplies to survive—and then refused to work anymore.

The Dutch saw all this turmoil and strife in England—and took their chance.

They put together a huge invasion fleet: 51 man-of-war ships, 3 frigates, 6 armed yachts, and other smaller vessels. This fleet was led by the famous Dutch admiral Michiel de Ruyter.

The Dutch also added 14 fireboats.

Fireboats were terrifying vessels. Older ships, filled with wood and soaked in pitch, were set on fire and sent crashing into enemy ships. Think of them as floating firebombs and ancient flaming torpedoes.

Example of a Dutch fireboat from the 1600s.

Even though England was in pretty bad shape, the Dutch realized this invasion wouldn't be an easy mission. They needed careful planning and precision.

The Dutch decided to attack through the River Thames, which led into the River Medway, where many English warships were being repaired at the Chatham Dockyard.

Both the Thames and the Medway are estuaries—wide rivers where fresh water mixes with the sea.

The Dutch hoped to destroy the dockyard at Chatham. That way, they could cripple the Royal Navy and wreck its ability to fix any damaged ships that survived the attack.

The English, meanwhile, were confident the Dutch would never dare to invade their homeland.

But on June 9, 1667, the Dutch fleet sailed up the Thames. The English tried to stop them on the way—by sinking their own ships and blocking the river.

But that tactic didn't work.

The Dutch ships kept coming.

England's King Charles II suddenly realized what a huge mistake it was not to have his navy ready. Now he panicked.

On June 10, the English shifted focus to Fort Sheerness, which protected the entrance to the Medway. If the Dutch captured this fort, they could destroy England's first line of defense, leaving the Medway wide open for further invasion.

About 300 English soldiers were garrisoned at Fort Sheerness. Unfortunately, they were poorly equipped. The Dutch opened fire with broadsides on the fort. Cannons belching flame and thunder, they bombarded the fort for two hours.

The English soldiers gave up and ran away.

Illustration of Dutch ships firing on Fort Sheerness.

Now the Dutch had secured their first victory—and gained a base for operations. After landing forces, they took supplies and weapons from the fort.

On June 11, the Dutch sailed up the River Medway.

Once again, the English sank more of their own ships to try to block them. But they also set up a tough defense. They placed three warships—the *Unity*, *Charles V*, and *Matthias*—and a giant iron chain across the river. This chain weighed 14 tons! Heavy enough to stop ships from sailing through, or they'd risk getting smashed into pieces.

The Dutch responded by sending in the fireboats first. The flaming ships charged at the English defenses, lighting up the sky like fire-breathing dragons.

The next day, because the river narrowed at this location, the Dutch sailed in a single-file formation and sent the smaller boats with explosives to attack the chain and the English defenses.

The English fired back with cannons placed on the riverbanks and from the ship *Unity*. But it wasn't enough to stop the Dutch.

Samuel Pepys feared the worst. It wasn't just the outcome of this battle that worried him. "And the truth is," he wrote, "I do fear so much that the whole kingdom is undone."

The Dutch kept pushing forward. Captain Jan van Brakel led the fireboats alongside a warship named *Vrede* with 40 cannons. They attacked *Unity* and captured it.

Anonymous 1667 Dutch painting, *Breaking the Chain Across the Medway.*

Dutch engineers jumped into the churning waters and began attacking the chain with hammers. Some reports claimed that fireboats were also sent in to explode on top of the chain and break it apart.

The Dutch then set fire to the English ships *Matthias* and *Charles V*.

In desperation, the English sank sixteen more of their own ships.

The Dutch rolled on, reaching Chatham Dockyard. They quickly destroyed the warships *Royal James*, *Loyal London*, and *Royal Oak*. Flames and smoke billowed across the river.

The Raid on the Medway, showing the Dutch fleet sailing up the River Medway toward Chatham, England. By Dutch painter Willem Schellinks (1623–1678).

Any remaining English ships weren't even fighting anymore—the sailors were too busy picking men out of the water to prevent their drowning.

Then came the ultimate national insult: the Dutch captured the *Royal Charles*, a ship that was the pride of the Royal Navy. The Dutch didn't set it on fire; they towed it away to Holland as a war trophy. A real insult!

On June 13 and 14, the Dutch finally sailed away. They had won the Battle of Medway—big time. They came, they conquered, and now they left the Royal Navy in complete disarray and humiliation.

The loss of this epic sea battle forced England into peace negotiations. Soon after, the Treaty of Breda was signed. As victors in this battle, the Dutch got the better end of the treaty, giving them the upper hand in world trade.

The battle also cost the English thirteen major warships, along with massive damage to the dockyard. And it was a national embarrassment that the Dutch towed away the *Royal Charles* as a war trophy.

However, the Battle of Medway spurred reforms concerning England's naval defenses. Among the changes were fortifying the Medway against enemy invasion and improving the Royal Navy fleet's readiness for battle.

And, within 100 years, England would become the strongest sea power in the entire world. Quite the comeback!

What about the Dutch navy?

Well, by the early 1700s, the Dutch navy was slipping into decline. By the 1800s, it was no longer among the world's great navies.

Sometimes a loss in battle turns into a greater win. And a win can a greater loss.

WHO FOUGHT?

Michiel de Ruyter.

Born in 1607, Michiel de Ruyter stands among the most famous naval heroes in Dutch history.

Known for his bravery and strategic thinking, he served as an admiral in the Dutch navy when Holland was fighting to protect its independence and defend its sea trade routes around the world.

De Ruyter is best known for leading bold and daring battles against England and France. The raid on Medway is considered among his greatest achievements. The victory elevated him to the status of national hero among the Dutch.

Despite his fierceness in these battles, De Ruyter was known for treating his crew and captains with respect, and for putting country above self.

He continued fighting for his country until he was 70 years old. However, De Ruyter was fatally wounded by a cannonball in a battle off the island of Sicily in 1676. He died a week later. His body was brought back to Holland, where he was given a state funeral. His nickname, *Bestevaêr*, is still used by the Dutch to this day. The name means "Grandfather."

INTERNET

A short documentary on the Battle of Medway: https://www.youtube.com/watch?v=3Zi3kYbVdWw

A longer video of the battle, and the "official video" of the Medway Council: https://youtu.be/6Z9l7y3LLe0?si=MHdTHE_I4c9ElM0g

The Battle of Lepanto

October 7, 1571

Battle of Lepanto by artist Jan Peeters.

Let's jump ahead to the year 1571—a time when the mighty Ottoman Empire (whose warriors were known as the Turks) was growing stronger and more powerful.

Spreading fear and chaos across the medieval world, the Ottoman forces raided coastal towns throughout Syria, Egypt, and Tunisia. They also took control of vast parts of Europe, including the Balkans and the cities of Buda and Pest (later joined to become Budapest, the capital of Hungary).

Now, as the Turks gained control over the Mediterranean Sea, they closed in on their next target: Rome.

At the same time, the Rome-based Catholic Church—also called Christendom—was dealing with other big problems. People were accusing church leaders of corruption, and a movement called the Protestant Reformation was challenging the Pope's authority.

With the Christians arguing among themselves, the Ottoman Empire continued to gain strength. Many Western Europeans worried that all of Western civilization was about to fall to the Ottomans.

Turkish helmet from the Battle of Lepanto.

To fight back, and regain control of the Mediterranean Sea, Pope Pius V ("Pie-us the Fifth") created a powerful alliance called the Holy League.

The League included Spain, several small Italian states (because Italy wasn't a single country yet), and the Knights of Malta—a brave group of warrior-monks who fought in battles and helped the sick, much like the more famous Knights Templar.

To fight the Ottoman navy, the Holy League assembled a fleet of warships and chose a man named John of Austria to lead it.

The Holy League fleet had 206 galleys—long, narrow ships powered by sails and oars. Some of the galleys had strong, sharp projectiles in the bows to better smash enemy ships and snap off their oars.

Ali Pasha.

The League also had six huge ships called galleasses, which were like floating fortresses armed with 30 cannons. Stretching some 160 feet in length, the galleasses also carried archers and musketeers, sailors who fired muskets at the enemy.

However, the Ottoman Empire had even more ships—222 galleys and 56 smaller ships called galliots. This fleet was under the command of a powerful admiral named Ali Pasha.

In early October, the Holy League fleet sailed out of the island of Sicily, today's Italy, to confront the Ottoman ships.

The two forces would meet in the Gulf of Patras, in the Ionian Sea, west of modern-day Greece. Nearby is the city of Lepanto, which gives this epic sea battle its name.

Some 65,000 men were aboard the Holy League ships. Half were soldiers. The other half were oarsmen—imagine that, more than 30,000 rowers!

Portrait of Pope Pius V, 1566, by artist Palma il Giovane.

Pope Pius also dispatched 30 priests on this mission. These Catholic chaplains were tasked with reminding the men that they weren't just fighting another battle. They were going to war to protect Christian civilization from falling to Islamic rule.

The Ottoman fleet also had more crew—about 50,000 sailors and oarsmen plus more than 30,000 additional soldiers. Among those soldiers were the Janissaries who, as children, were kidnapped by Muslim warriors, forced to convert to Islam, and trained into elite soldiers. These Janissaries were known for unquestioning obedience to superiors, deadly use of firearms, and fatal fighting skills in battle.

In the hours before this battle broke out, John of Austria stood on his ship *The Real* and gave a speech to his men. As he spoke, he held up a crucifix—a cross with Jesus on it.

"My children," said John of Austria, "you've come to fight the battle of the cross—to conquer or to die. But whether you are to die or conquer, do your duty this day and you will secure a glorious immortality."

On the morning of Sunday, October 7, a hard-blowing wind tufted the sea with choppy waves. A sailor perched high on a Holy League ship suddenly spotted the Ottoman sails approaching in the distance.

Quickly, both sides moved into battle formation.

This fresco from the late 1500s depicts the Battle of Lepanto between the Christian force and the Ottoman navy. Notice the ships in reserve while the clash takes place in the middle. This fresco can be seen in the Hall of Maps in the Vatican Museums in Rome.

With more ships, the Ottoman line was longer than the Holy League's—about 3,000 feet longer—with ships arranged in squadrons, or military units. Each squadron had about 60 galleys. The Ottomans also held a small reserve unit behind the main line.

The Holy League's formation placed four galleasses in front, with John of Austria leading the fleet's center with 62 galleys. On his right and left were 53 galleys each. In reserve, he kept 38 more galleys and 2 galleasses.

Imagine this scene.

More than 450 ships carrying 100,000 sailors and 60,000 soldiers ready to inflict deadly harm on each other while the wind whipped the sails and oarsmen rowed to a steady hammering beat below deck. Both sides arrayed their power, preparing to sink the enemy ships and slaughter the crews.

The Ottoman fleet also approached with noise—drums and ululations (a shrieking yodel)—that was supposed to put the fear of God into their enemies.

But the Holy League galleys and galleasses were silent. John of Austria had told them not to fire until they were close enough "to be sprayed by Muslim blood." The League crews were also spreading grease at points on the deck. If the enemy tried to board, they would slip and fall.

Ali Pasha was leading the Ottoman fleet's center with 87 galleys. His right wing contained 60 galleys and 2 of the smaller galliots. His left wing held 61 galleys and 32 galliots. In reserve behind those ships were 8 galleys and 22 galliots.

At first, that hard-whipping wind worked against the Holy League ships, making it harder for them to steer against the force.

But the wind suddenly shifted. Now the two flagships—the *Real* with John of Austria and the Ottoman *Sultana* with Ali Pasha—were heading straight for each other!

John of Austria in battle at the bow of his ship in the Battle of Lepanto. Cannons roared and blasted apart sails. Ramming bows smashed into the ships, spewing deadly wooden splinters that flew across the decks. Musketeers fired from behind cover while arrows flew overhead and zinged through the smoke-filled air. Grappling hooks clanked and lashed the enemy ships together. Soldiers leapt over the sides, cutlasses and sabers ready for brutal hand-to-hand combat.

The *Real* and the *Sultana* were locked in a deadly duel, one that looked like John of Austria was losing. Twice the Turkish warriors nearly boarded the *Real,* a battle-ending potential since the *Sultana* was carrying about 300 Janissaries and many musketeers.

Then suddenly—BOOM!

The Holy League ship, the *Capitana,* rammed the *Sultana* and unleashed a storm of musket fire.

One artist's depiction in 1574 of the Christian and Ottoman forces lined up in the Battle of Lepanto. In the lower portion, the ship with the white banner is the *Capitano*, ramming the *Sultana*.

Seizing the moment, The *Real*'s soldiers jumped aboard the *Sultana*. Amid the eruption of chaos, Ali Pasha was shot in the forehead and killed instantly. Within moments, the Ottoman flag—a crescent moon and star—was yanked down. The Christian banner was raised.

Meanwhile, the Holy League's gigantic galleasses continued hammering the Ottoman fleet, whose sailors now saw the new flag raised on the *Sultana*. They knew their leader was gone. Without Ali Pasha's orders for what action to take next, the Ottoman ships scattered in a confused manner.

Smoke and fire filled the air as the sea turned red with blood. Bodies floated on the water.

One artist's view of the Battle of Lepanto.

The Ottoman ships that could still navigate the waters turned and fled the battlefield.

By 4 p.m. that day, the Holy League declared victory in this epic sea battle.

Sixty Ottoman ships were beached. Fifty-three had sunk. Another 117 were captured. About 30,000 sailors and soldiers were killed or taken prisoner. Many of those prisoners would now row the Holy League ships, as the slaves—unchained to fight in the battle—were now set free. Some estimates are that 12,000 slaves were set free in this battle.

The Holy League lost 13 ships. More than 7,000 men had died, and another 8,000 were wounded.

Moreover, the Battle of Lepanto effectively ended Turkish control of the Mediterranean Sea. The victory also stopped the Ottoman Empire from expanding further into Europe.

Many historians consider the Battle of Lepanto to be the greatest naval battle of the Middle Ages.

Militarily, the Battle of Lepanto marked the last major engagement in the Western world fought almost entirely between rowing vessels—those galleys and galleasses. From then on, oar-powered ships were phased out in favor of sails, which offered better maneuverability.

Morally, the battle encouraged Christian Europe. The Ottomans no longer seemed like an unbeatable force. And from this point onward, Ottoman power began to wane. Within a century, it was no longer a threat to Christendom.

WHO FOUGHT?

Knights of Malta.

In the 11th century (about 1,000 years ago), a group of monks known as the Knights Hospitaller took care of sick people in the holy city of Jerusalem.

These monks also helped protect pilgrims who traveled to such holy places to worship Jesus Christ.

But they soon needed to become battle-trained knights to defend Christians against pirates, armies, and enemies of the Christian faith.

In 1530, for their own safety, the Knights Hospitaller moved to an island in the Mediterranean Sea called Malta. At that point, the Hospitallers became known as the Knights of Malta. And their mission grew to protect Europe from pirates and Ottoman invaders.

The knights wore special armor adorned with a white cross on the chest. For hundreds of years, they helped keep peace. They also became famous for their courage in battle and their kindness to people in need, especially the sick and the poor.

Incredibly, the Knights of Malta still exist today. The group is much smaller—about 13,000 members—than in the days of castles, kings, and armies. Also, the knights no longer fight battles.

Headquartered in Rome, Italy, the knights' main purpose today is helping needy people around the world by maintaining hospitals, offering medical care, and aiding with disaster relief.

While these knights no longer use swords and armor, they still wear that special white cross on their chest to show their dedication to Jesus Christ and helping the needy.

BOOKS

Medieval Castles and Knights for Kids: An Enthralling Guide to Strong Fortresses and Brave Heroes of the Middle Ages by Billy Wellman

INTERNET

Jake Phillips, who narrates some the *Great Battles for Boys* audiobooks, reads the poem "Lepanto" by G.K. Chesterton on his

YouTube channel, The Cultured Bumpkin. Listen to the poem's stirring lines, such as, "And the Pope has cast his arms abroad for agony and loss, And called the kings of Christendom for swords about the Cross …"

Listen to the whole poem here:

History Simplified has a visual depiction of the Battle of Lepanto: https://www.youtube.com/watch?v=QuYQkIQWLHk

The Daring Raid on Tripoli

February 16, 1804

Burning of the Frigate Philadelphia *in the Harbor of Tripoli, February 16, 1804,* oil on canvas, by Edward Moran (1829–1901).

WE'RE NOW GOING TO LEAP FORWARD more than 200 years to the 1800s.

By this time, the Ottoman Empire had shrunk—significantly—in both size and power.

However, a dangerous group of Ottoman-backed pirates still thrived in the Mediterranean Sea. These bandits were known as the Barbary pirates.

They kept a sort of headquarters in Tripoli, in North Africa.

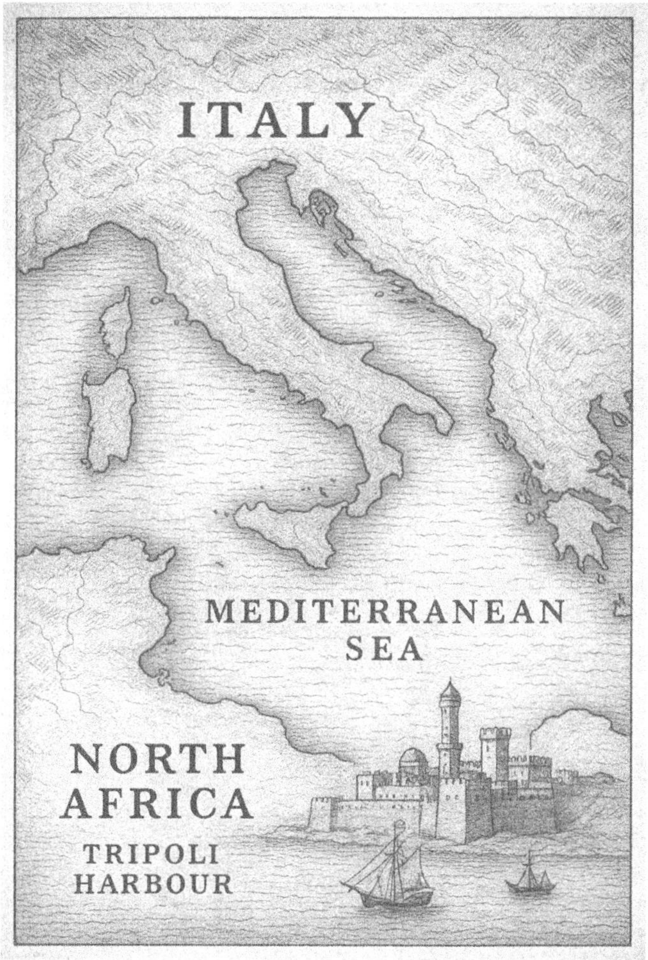

The Barbary pirates sailing the Mediterranean Sea would capture American merchant ships, steal their cargo, and kidnap the sailors—then demand huge ransoms for the sailors' release, or else the sailors would be sold into slavery.

The pirates called these ransom demands "tributes." Basically, the deal was: pay us enough money and we'll stop kidnapping your sailors and terrorizing your ships.

When U.S. President Thomas Jefferson refused to pay the tributes, the Barbary pirates declared war on the United States.

The United States didn't want a war. It was still a very young country at this time, and it had just finished fighting an all-out war against Great Britain in order to gain its independence (you can read about that war in my book on the American Revolution).

Painting of a Barbary pirate, by Pier Francesco Mola, 1650.

Instead of going to war, President Jefferson set up a naval blockade—which is kind of like posting a "No Entry" sign on the ocean. American ships would block hostile vessels from entering or leaving.

Although the U.S. Navy was quite small at the time, Jefferson sent a fleet of warships to patrol the Mediterranean Sea and prevent pirate ships from attacking or capturing American merchant vessels.

The goal was to weaken the Barbary pirates' power.

But the pirates were tricky. They had their own fleet of ships, which often hid in ports. The pirates also disguised the ships as ordinary merchant vessels and managed to sneak past the American blockade.

And then, disaster struck.

In October 1803, an American ship, the USS *Philadelphia*, ran aground near Tripoli's harbor. The Barbary pirates captured the ship and imprisoned its sailors.

Now, the pirates not only had possession of a powerful U.S. warship but also planned to copy its design to build their own navy.

The United States needed to act fast.

Among the officers originally assigned to the naval blockade was a young lieutenant named Stephen Decatur, who had captained the *Philadelphia* prior to its capture.

CAPTAIN STEPHEN DECATUR, JUN., U.S.N.

(From A. B. Durand's engraving of the portrait by T. Sully.)

Captain Stephen Decatur.

Hearing that the *Philadelphia* had been taken by the pirates, Decatur began hand-picking a crew of volunteers for the extremely dangerous mission of retaking the *Philadelphia*—inside enemy territory!

Eventually, Decatur selected about seventy-five sailors and set out on the mission.

For days, the crew battled lashing storms until the night of February 16, 1804, when they sailed a small, two-masted ship called the *Intrepid* into Tripoli Harbor.

But then, a light wind slowed their approach, creating a new hazard. More than 100 cannon batteries protected the harbor and the *Philadelphia*. Ottoman guards stood watch all around.

Suddenly, an Ottoman lookout aboard the *Philadelphia* spotted the *Intrepid* approaching. He called out in Arabic, demanding to know the crew's purpose.

Cleverly, Decatur had planned for this.

His crew was disguised as traders from the island of Malta, and the *Intrepid*'s pilot spoke enough Arabic to respond. He explained that they had lost their ship's anchors in the recent storms—could they please tie up their vessel to the large, anchored ship for the night?

The guard agreed.

But just as the *Intrepid* was about to tie up to the *Philadelphia*, the wind shifted, pushing the *Intrepid* about 20 yards away from the warship.

Later, one of Decatur's crew described those seconds as "a moment of great anxiety."

Decatur quickly ordered his men to adapt. They climbed aboard a smaller vessel they were towing and rowed it to the *Philadelphia* to tie a line to it.

Boarding the Philadelphia, by J.O. Davidson.

Just as that task was about to be completed, one of the Tripolitan guards realized why these men were wielding cutlasses and axes.

The guard yelled for help—but it was too late.

Decatur and another sailor jumped onto the *Philadelphia*. More crew followed. And hand-to-hand combat broke out, with the sailors slashing their way down the *Philadelphia*'s deck.

The Americans took control of the ship.

But their cover was blown.

Ottomans were yelling from shore, firing arrows, and blasting cannons at the Americans.

The raiders scrambled below deck but found no one there. And it was now clear that during its time in pirate hands, the *Philadelphia* had fallen into serious disrepair.

Decatur now realized the warship couldn't be towed away—and the risk wasn't worth the attempt.

Hearts pounding and swords ready, the men obeyed Decatur's orders and placed explosive charges throughout the ship—better to destroy the mighty vessel than allow the enemy to keep it and copy its design.

The fire spread quickly. Sparks burst across the wooden deck. Flames plumed into the night sky over Tripoli Harbor. The *Philadelphia*'s wooden masts suddenly looked like giant torches, spewing fire in every direction.

The sudden fire further shocked the enemy—what was happening?! Screams erupted into the night. Men aboard nearby vessels leapt into the water to escape death by incineration.

And then: KA-BOOM!

The fire reached the ship's magazine, where gunpowder was stored. The *Philadelphia* exploded with a final giant blast.

By then, Decatur and his men were already back on the *Intrepid* and sailing out of Tripoli Harbor.

Within about 10 minutes they had completed this daring mission—and the brave officer who led it hadn't lost a single man.

Now the crew looked back across the inky black water and watched the *Philadelphia* sink into the depths of Tripoli Harbor.

While this raid isn't technically an epic sea battle, it remains a highly crucial event in United States naval history.

First, the daring raid sent a powerful message to the Barbary pirates—and to the world: America might be a young country, and its navy quite small, but it was not afraid to fight back—within enemy territory—and use any means necessary.

Second, while the initial blockade was one of the first real tests for the young U.S. Navy, the raid marked a major turning point in how America would handle foreign threats.

Jefferson's refusal to pay tribute, and his decision to launch a naval blockade and bold raid instead, showed iron-willed leadership and clever decision-making by the nation's commander in chief.

Following the mission, early America earned more respect on the world stage.

Even British Admiral Horatio Nelson, considered among the greatest naval commanders of all time, called the raid on Tripoli Harbor "the most daring act of the age."

WHO FOUGHT?

Stephen Decatur was born in the midst of the American Revolution, on January 5, 1779.

His family lived on the Eastern Shore of Maryland. His father, Stephen Decatur Sr., was a commodore in the Continental navy during the Revolutionary War. Decatur Sr. raised young Stephen in a nautical world of ships and sailing.

After college, Stephen followed in his father's footsteps and joined the U.S. Navy.

He was 19 years old.

Six years later, he was promoted and became the youngest man to reach the rank of captain in the history of the U.S. Navy.

He served under three presidents:

1. Thomas Jefferson, during the First Barbary War (1801–1805), including his famous 1804 raid to destroy the USS *Philadelphia* in Tripoli Harbor.
2. James Madison, during the War of 1812. Decatur commanded ships such as the USS *United States* and fought in significant naval engagements.
3. James Monroe, in the postwar period, including the Second Barbary War (1815), when Decatur led a successful mission against Algiers and helped negotiate peace treaties with several Barbary states.

Decatur played a major role in the early development of the U.S. Navy. His reputation grew due to his natural ability to lead and for his genuine concern for the seamen under his command.

In 1818, he built a large home in Washington, DC, on Lafayette Square, which still stands today. Known as Decatur House, it became a hub for Washington society and for his close personal friends, such as President James Monroe.

In 1820, Decatur's life and career came to a sudden end when he refused to take back some remarks he'd made about

Commodore James Barron, who had surrendered to the British in a naval battle.

Barron challenged Decatur to a duel.

Decatur lost the sword fight.

But his reputation as a national hero lives on.

BOOKS

Stephen Decatur: The Life and Legacy of the Youngest Navy Captain in American History (Charles Rivers Editors)

The Barbary Pirates 15th–17th Centuries by Angus Konstam

The Barbary Pirates by C.S. Forester

C.S. Forester, creator of the beloved Horatio Hornblower series, takes young readers on an exciting semi-fictional adventure to the shores of Tripoli, when the United States was threatened by pirates who snatched American merchant ships and imprisoned sailors.

INTERNET

The official hymn of the United States Marine Corps mentions the raid on Tripoli Harbor: "From the Halls of Montezuma / To the shores of Tripoli / We fight our country's battles / In the air, on land, and sea …"

You can hear the rest of the hymn here, along with historical naval footage: https://youtu.be/pqqhq63RfmA

MOVIES

Tripoli (1950)

The Battle of Trafalgar

October 21, 1805

Ships clashing in *The Battle of Trafalgar*, an 1839 engraving by artist William Miller.

FOR OUR NEXT EPIC SEA BATTLE, we're staying in the early years of the 1800s—but shifting over to Spain and the Atlantic Ocean, where a massive, world-changing naval clash was brewing on the high seas.

On one side of this epic confrontation stood British Admiral Horatio Nelson, the nation's greatest naval hero, commanding a formidable fleet of 27 warships.

On the other side were the combined naval forces of France and Spain—21 French ships under Admiral Pierre-Charles Villeneuve and 6 Spanish ships under Admiral Federico Gravina.

Nelson, recognizing the gravity—and the deadly possibilities—of the coming battle, offered a simple yet powerful prayer. In fact, it was almost a plea:

"May the great God ... grant to my country a great and glorious victory," Nelson said.

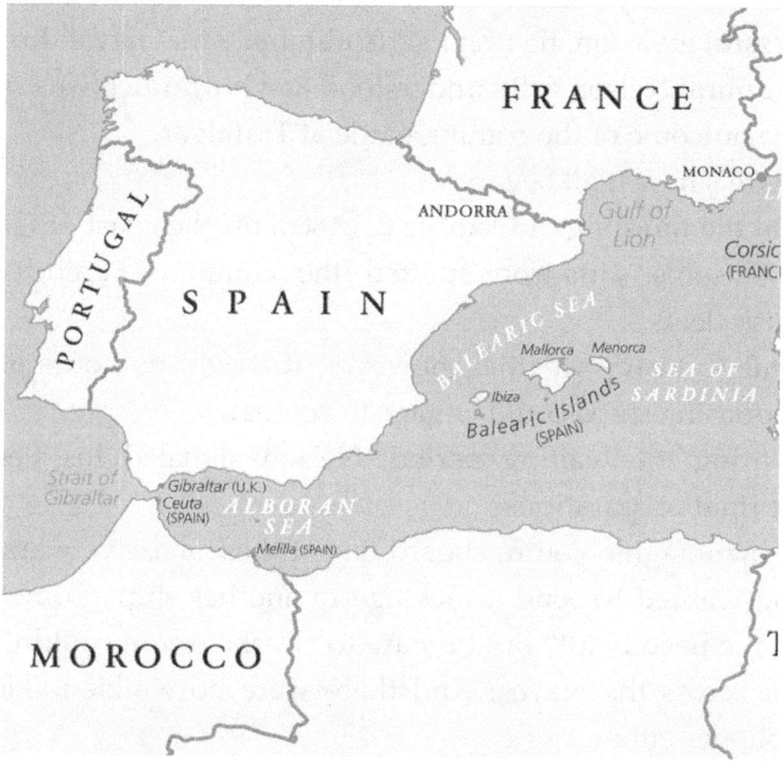

The Strait of Gibraltar sits between the southern tip of Spain and the northern tip of Morocco.

More than just a clash at sea between powerful naval forces, this battle was set to determine the fate of Europe.

How?

Well, if the French and Spanish fleets managed to defeat the British, the French emperor, Napoleon Bonaparte, planned to invade England. On land, England stood little chance against the fierce and brilliant French general.

Napoleon had already amassed an army of 167,000 troops and prepared over 2,000 gunboats and transport vessels to cross the English Channel from France.

But the British did have one major advantage: their navy.

Renowned for its dominance and string of victories at sea, the Royal Navy stood as Napoleon's greatest obstacle. For a successful invasion, he first had to eliminate that naval threat.

Admiral Nelson fully understood just how much was riding on the outcome of the coming battle at Trafalgar.

Here's how it all began.

On the morning of October 21, 1805, off the coast of Gibraltar, Nelson's squadron spotted the combined French and Spanish fleets.

Due to heavy weather, however, it would be hours before the opposing ships could engage in combat.

During the waiting period, Nelson signaled his fleet to make final preparations.

Now imagine you're aboard one of these massive warships. If you wanted to send a message to another ship—something like "We need help!" or "Prepare to attack!"—you couldn't just shout across the waves. And there were no walkie-talkies or cell phones either.

Instead, sailors used colorful flags to communicate. Each flag represented a letter, number, or sometimes an entire word. The crew would hoist them up in specific sequences on the ship's mast. It was a bit like sending a secret code—only those who understood the flag system could read the message.

An example of signal flags during the Age of Sail—the period from roughly the mid-1500s to the mid-1800s.

As the battle neared, the oceanic weather brooded.

From his flagship, the HMS *Victory*, Admiral Nelson ordered his signal officer to send up a message in flags. The coded colors spelled out:

"England expects that every man will do his duty."

The *Victory* was an impressive warship, armed with 100 guns and crewed by 850 men. Meanwhile, the French flagship, the *Bucentaure*, carried 80 guns, and the Spanish flagship, the *Santísima Trinidad*, boasted a staggering 136 guns—the largest ship in the battle.

Nelson was outnumbered in ships, firepower, and manpower.

The marines aboard these ships had a grim and focused task: to fire on enemy crews and eliminate as many men as possible. For that purpose, HMS *Victory* featured a three-tier gun deck, with rows of heavy cannons.

Each cannon sat on wooden wheels so it could be maneuvered into firing position. The ship's gunports—wooden shutters that opened outward—would swing open when engaging the enemy. After each blast, the force of the cannon's recoil caused it to roll backward. The gun crew would then quickly swab the barrel clean (to remove residual gunpowder that could cause a dangerous misfire) and reload for the next shot.

Also on board were mortars—smaller, high-trajectory cannons—used to rain fire down onto enemy decks.

Originally, French Admiral Villeneuve had planned to flee toward the Mediterranean Sea, avoiding battle altogether. The French command genuinely feared the might of the British fleet and the reputation of the legendary Nelson.

But Villeneuve changed his mind. Perhaps he feared being disgraced if he didn't engage. On this morning, he gave the order: turn the fleet about—prepare to attack the British.

Keep in mind, these were sailing ships. The speed of their approach and maneuvering was determined by the wind.

At 10 a.m., the Battle of Trafalgar began.

The Battle of Trafalgar, ships firing upon each other. By an unknown artist.

According to standard naval tactics of the time, fleets would position themselves in parallel battle lines facing each other. Then both sides would open fire—broadside, meaning all the guns on one side of the ship would fire at once.

But Admiral Nelson was legendary for many reasons—and this battle would add yet another to the list.

Facing two combined fleets with more ships, more guns, and a longer line of battle, Nelson knew the traditional approach wouldn't be enough.

So he changed the strategy.

Instead of matching the enemy line, Nelson would sail his fleet directly toward the enemy ships—attempting to cut right through their formation.

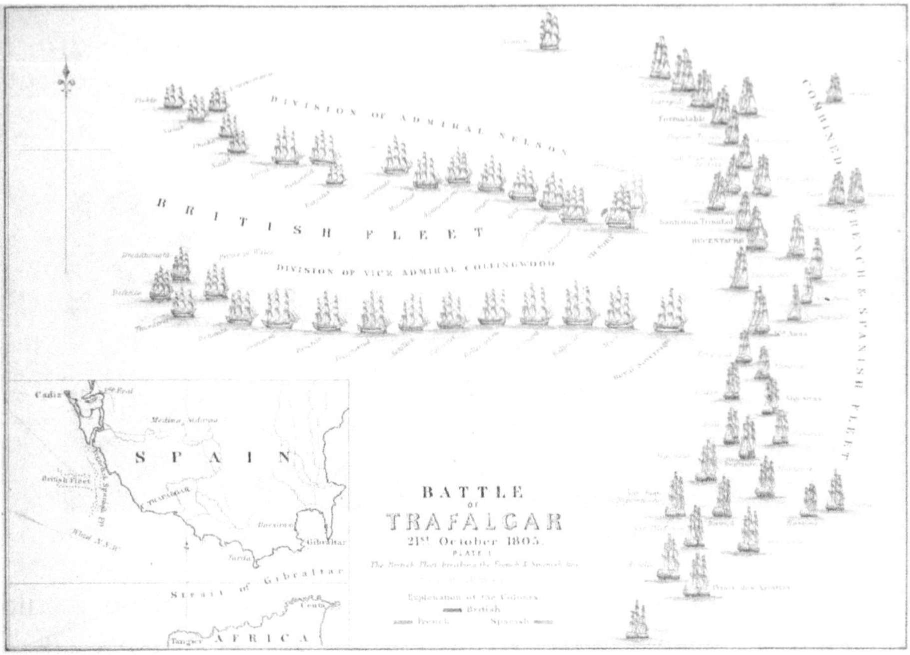

British fleet, left, sailing straight at the French and Spanish lines on the right.

In the opening salvos, or bombardment, Nelson's daring fleet took heavy cannon fire. The combined French and Spanish lines unleashed deadly broadsides against the approaching British ships.

Nelson's *Victory* came under relentless attack—including from the 136 guns of the massive *Santísima Trinidad*. Cannonballs smashed into the wooden hull and decks of the *Victory*, sending deadly splinters flying through the air like missiles. Sails were shredded, rigging snapped, and men fell where they stood—struck down by cannon and musket fire.

But then, amid the choking smoke and deafening roar of battle, Nelson's brilliant plan began to work.

By sailing his fleet directly into the flank of the enemy formation, he shattered the long allied line—breaking it into pieces.

The *Victory* breaking through the French line. By an unknown artist.

Villeneuve wasn't prepared for such a daring move. Nelson's columns continued their charge, splitting the Franco-Spanish fleet and ultimately severing it into three isolated sections.

The scattered enemy ships tried to regroup—some attempting to turn back and rejoin the fight—but now the British had gained a temporary numerical superiority. Nelson's tactic of breaking the line had paid off. By cutting into the center, he had effectively neutralized the firepower at either end of the enemy formation. Those ships simply couldn't reach the British forces that were now attacking the middle.

The only way the French and Spanish ships at the ends of the line could engage was by turning around and sailing back toward the center—an enormous challenge in the chaos of battle, and entirely dependent on wind speed and direction.

With cannon smoke darkening the sky, the *Victory* maneuvered behind the French flagship *Bucentaure*. Then, the British unleashed a devastating broadside—killing scores of men and destroying 20 of her cannons.

Nelson's other ships broke through the enemy lines as well. The brilliance of his plan was now fully revealed.

When the British fleet sailed through the French and Spanish formation, the enemy lost their advantage in numbers. The battle devolved into brutal, one-on-one, ship-to-ship combat.

Cannon smoke billowed into the October air. Ships crashed into one another. Explosions flashed. Death arrived at point-blank range—so close, it was nearly impossible to miss.

The British gunners had a decisive edge: they could load and fire their cannons more quickly and accurately. That speed mattered.

Soon, the *Victory* came alongside the French ship *Redoutable.* British gunners pounded her hull with cannon fire, while marines, armed with muskets, aimed at the rigging and enemy crew—killing nearly 50 men.

But *Redoutable* struck back with deadly force, killing a similar number aboard the *Victory.* Among the most seriously wounded was Admiral Nelson. A musket ball pierced his shoulder, passed through his lungs, and lodged in his spine.

Yet the battle raged on.

Nearby, the British ship *Temeraire* raked *Redoutable*'s deck with ferocious cannon fire—killing another 487 men. At last, *Redoutable* surrendered.

Next to fall was the mighty Spanish *Santísima Trinidad.* After suffering the loss of roughly 450 men, she struck her colors at 4:15 p.m., signaling surrender.

Santísima Trinidad *demasted after the Battle of Trafalgar*, by R.B. Spencer.

Meanwhile, the French Admiral Villeneuve, who was hoping for victory or death, received neither. He was forced to surrender his flag as well.

However, this great English victory brought a tremendous loss for the country.

At 4:30 p.m., Admiral Nelson died.

A painting depicting the death of Admiral Lord Nelson at the Battle of Trafalgar on October 21, 1805.

The victory at Trafalgar allowed Britain to take control of 18 French and Spanish ships.

The British hadn't lost even one in this daring battle.

In total, the French and Spanish lost 14,000 men who were either killed or captured. The British lost 1,500.

Moreover, the British victory ended Napoleon's plan to invade England.

For the time being, England could continue to rule the ocean waves—and itself.

WHO FOUGHT?

Admiral Lord Nelson.

In the annals of British history, Horatio Nelson stands among the most famous and admired naval commanders.

He was born in 1758 in the small town of Burnham Thorpe, in Norfolk, England. At just 12 years old, he joined the Royal Navy as a midshipman, sailing under the command of his uncle.

From the start, Nelson showed dedication, courage, and a deep understanding of naval warfare. His talent earned him rapid promotions, and he became a captain by the age of 20—an unusually young age for such a responsibility. Eventually, he rose to the rank of vice admiral and became one of Britain's most respected military leaders.

As a commander, Nelson was both brilliant and fearless. He fought in key battles around the globe—from the Caribbean to the Mediterranean, from the Indian Ocean to the coasts of Europe. He gained fame for his bold tactics, his personal bravery, and his willingness to share danger alongside his men. His leadership inspired fierce loyalty.

Nelson was wounded multiple times in battle. He lost sight in one eye during a campaign in Corsica and later lost his right arm in a failed attack on the Spanish port of Santa Cruz de Tenerife. Despite these serious injuries, he refused to retire from service and continued to lead his fleet.

His most legendary moment came during the Battle of Trafalgar when Nelson made that bold and unconventional move of sailing directly into the enemy line—splitting it and throwing the opposition into chaos. His strategy not only led to one of the most decisive naval victories in British history but also crushed Napoleon's hopes of invading Britain.

It's not an exaggeration to say Nelson saved his country—and died for it after being mortally wounded during the battle by a French sniper aboard the *Redoutable*.

But before he died, Nelson learned that the battle had been won. His final words were, "Thank God, I have done my duty."

Today, Lord Nelson is remembered as a national hero. In the heart of London, a towering column crowned with his statue stands in Trafalgar Square—built to honor the man whose courage and leadership helped shape the course of history.

BOOKS

Young Nelsons: Boy Sailors During the Napoleonic Wars by D.A.B. Ronald

The Battle of Trafalgar 1805: Profile Models of Every Ship in Both Fleets by Florian Richter (author), Peter Dennis (illustrator). Presents color profile models of every ship on both sides in this epic battle, which can be cut straight out of the book to create the British, French, and Spanish fleets.

INTERNET

A short, animated explanation of the battle: https://www.youtube.com/watch?v=YeyTC9vTtLc

Epic History presents Nelson's many battles in 3D form, and gives historical context: https://www.youtube.com/watch?v=FV-0f4NfQ60

The Battle of Manila Bay

May 1, 1898

"Remember the *Maine*"

Battle of Manila Bay, by James Gale Tyler (1898).

IN THE LATE 1800S, Spain still controlled several colonies around the world. A colony is a land ruled by a foreign country, usually located far away.

One of those Spanish colonies was the island of Cuba, located just 90 miles off the southern coast of Florida—so close that on a clear day, you can see Cuba from the southern tip of the state.

As close neighbors, the United States had many business ties with Cuba, especially in the sugar trade. The Cuban people had taken note of how the United States had once been a colony of Britain but had fought for—and won—its independence.

Now, the Cubans wanted their own freedom from Spanish rule.

The Spanish government, meanwhile, often treated Cuban revolutionaries and civilians harshly. Reports described Spanish soldiers burning villages, imprisoning rebels without trial, and forcing civilians into crowded camps—where many died of disease and starvation. This brutal policy, known as "reconcentration," was widely condemned.

Public sympathy in the United States began to grow for the Cuban cause.

In 1897, U.S. President William McKinley addressed Congress and said that Spain should be given a chance to improve its governance in Cuba and negotiate a peaceful resolution. But as a show of support for the Cuban people—and to protect American interests and citizens on the island—McKinley ordered the battleship USS *Maine* to sail to Cuba and anchor in Havana Harbor.

Though often described as a "war hawk"—someone eager to start combat operations—McKinley was actually cautious about entering a war with Spain. Within Congress, however, opinion was divided. Some members—true war hawks—wanted to invade Cuba, not just to help the Cuban revolution but to take control of the island and its lucrative sugar plantations. Others in Congress believed the U.S. should support Cuban independence but not seize the island as a colony.

But everything changed on the evening of February 15, 1898.

KA-BOOM!

A massive explosion tore through the USS *Maine*. The ship quickly sank to the bottom of Havana Harbor, killing more than 250 American sailors.

Who had caused the explosion?

Spanish forces?

Cuban rebels?

Or was it an accident—maybe a coal fire igniting the ship's ammunition?

The truth remains unclear to this day.

Modern investigations suggest a coal bunker fire could have accidentally triggered the blast, but at the time, no conclusive evidence existed.

Still, the American public and press were furious.

Newspapers—especially those owned by William Randolph Hearst and Joseph Pulitzer—ran dramatic and emotional headlines blaming Spain, helping to stir public outrage. (This style of sensational reporting became known as "yellow journalism.")

Those stories, however, launched a phrase that echoed across the country:

"Remember the *Maine*! To Hell with Spain!"

The sinking of the *Maine* was the spark that would ignite an American passion for war with Spain, and ultimately lead to a new era of the U.S. as an emerging world power.

This illustration, titled "And Boys, Remember the Maine!," by cartoonist Clifford Berryman, which appeared in *The Washington Post* on April 3, 1898, depicts an angry Uncle Sam addressing sailors as the USS *Maine* sinks in the background.

Naturally, after the explosion of the *Maine,* tensions between the United States and Spain escalated sharply.

Spain didn't want America meddling in what it considered a colonial issue. Meanwhile, Americans—outraged over the deaths of the 250 sailors—wanted justice. Or, perhaps more accurately, revenge.

Diplomatic negotiations continued for several tense weeks. President McKinley still hoped to avoid war, urging Spain to grant Cuba independence or at least some meaningful reforms. But Spain refused, unwilling to lose control of one of its last remaining colonies.

Eventually, the talks failed.

On April 24, 1898, Spain formally declared war on the United States.

The very next day, April 25, the U.S. Congress responded with a declaration of its own—and then backdated it to April 21, three days earlier. It was a political move, almost like saying: "Oh really? Well, we declared war first!"

(Sounds petty, doesn't it? But in moments like these, political theater gets intense—heated enough, unfortunately, to spark actual warfare.)

At the time, the Assistant Secretary of the Navy was an up-and-coming political figure named Theodore "Teddy" Roosevelt. One of the strongest war hawks in Washington, Roosevelt believed that war with Spain would not only help liberate Cuba but also elevate the United States to the status of a global power.

Roosevelt didn't just advocate for war from behind a desk. He was about to do something bold: he would personally join the fight.

Teddy Roosevelt in 1898.

However, neither the United States nor Spain was truly prepared for a full-scale war.

The U.S. Army was small—only about 25,000 men—and underfunded. Most American troops were stationed in the western territories of the United States, still engaged in what were known at the time as the "Indian Wars." The army was trained more for small skirmishes and ambushes on the frontier, not for large-scale battlefield engagements against a formal European-style army like Spain's.

To make matters worse, the U.S. Army was still using some outdated weapons, including old muskets and antiquated rifles. In summary: supplies were short, training was inconsistent, and the army lacked the organization needed for a complex overseas campaign.

But even more troubling was the lack of adequate medical supplies and almost no understanding of tropical diseases. Cuba's hot, humid climate was a breeding ground for deadly illnesses such as yellow fever and malaria, which would later claim more American lives than actual combat.

And yet another problem? Transportation. The United States simply didn't have the infrastructure—basic facilities and systems—to quickly and efficiently move an entire army across the Gulf of Mexico and onto Cuban soil.

However, the U.S. did have one major advantage: a modern navy. And that was in large part thanks to that same Assistant Secretary of the Navy, Theodore Roosevelt.

Roosevelt had long recognized that a powerful navy would be key not just for this war, but for America's future on the world stage. While other world powers—like Britain, France, and Germany—already had large and advanced fleets, the U.S. had been lagging behind. Roosevelt helped change that.

He championed the use of new, state-of-the-art battleships, including the USS *Indiana*, USS *Iowa*, USS *Massachusetts*, and USS *Oregon*. These powerful ships formed the core of America's newly energized Atlantic fleet. But the ships were scattered across the ocean at the outbreak of war, making coordination difficult.

In addition to these battleships, the U.S. Navy also had several modern cruisers—vessels smaller than battleships but built with steel hulls, improved armor, stronger engines, and modern gun turrets. These cruisers made up part of what was called the Asiatic Squadron, stationed in Hong Kong, far from the conflict in the Caribbean.

By the way, it's important to note that a fleet refers to a large collection of warships under a single command with broad strategic roles. A squadron is a smaller, more focused group of ships, often organized around specific missions or geographic regions.

The Asiatic Squadron was under the command of Commodore George Dewey, who oversaw a powerful group of ships: the cruisers USS *Olympia* (his flagship), *Baltimore*, *Boston*, and *Raleigh*, along with two smaller gunboats.

And while the world's eyes were on Cuba, Dewey and his squadron were about to launch a surprise attack that would change the course of the war—and history.

The USS *Olympia*, Commodore Dewey's flagship.

Gunboats are small, lightly armored ships equipped with medium- and large-caliber guns. They're specifically designed to operate in shallow coastal waters, rivers, and harbors—places where larger warships can't safely maneuver.

In addition to two gunboats, Commodore George Dewey also had one cutter—a small, fast, agile vessel useful for patrol or reconnaissance—and two cargo ships to resupply the fleet.

In command of the Spanish fleet in the Philippines was Admiral Patricio Montojo y Pasarón.

Spain's army was far larger than America's—boasting around 200,000 soldiers, nearly ten times the U.S. total. But its navy in the Pacific was in poor shape. Spain had spent years suppressing rebellions across its colonies, including in Cuba, the Philippines, and Puerto Rico. These drawn-out conflicts, combined with a struggling economy, had left much of the Spanish naval fleet outdated, under-armed, and in need of serious repairs. Many of their ships still had wooden hulls, limited armor, and obsolete weapons.

But, on paper, the Spanish fleet in the Philippines outnumbered Dewey's squadron.

Spain had about 40 vessels, though many were small or auxiliary ships. Their main warships included the cruisers *Reina Cristina, Castilla, Isla de Cuba, Isla de Luzón, Don Juan de Austria,* and *Don Antonio de Ulloa.*

While these ships looked impressive in formation, they were no match for America's newer, faster, and better-armed steel warships.

Right after Congress officially declared war on Spain, the U.S. Navy Department sent a clear and urgent message to Commodore Dewey in Hong Kong.

Here's part of what that message said:

"Proceed at once to the Philippine Islands. Commence operations . . . against the Spanish fleet. You must capture vessels or destroy."

And with that order, Dewey and his squadron set sail for the Philippines—headed straight for the first major naval battle of the Spanish-American War.

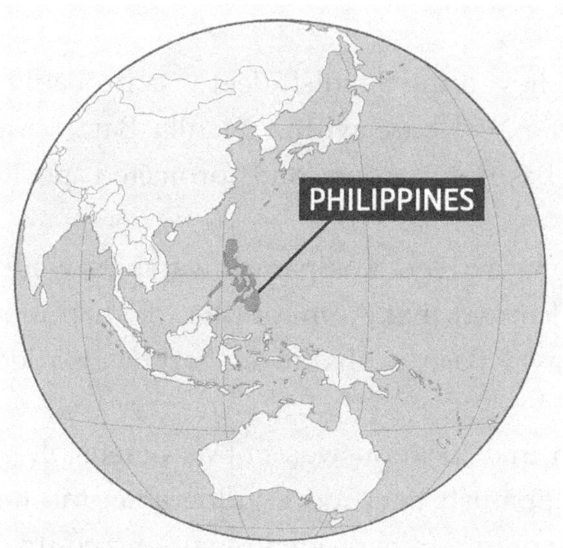

Global location of the Philippine Islands. The larger countries around it are China, above, and Australia, furthest below.

Why was the United States sending Commodore Dewey to the Philippines?

Because those islands were another colony of Spain. So while the war may have started over Cuba, it quickly became a global conflict.

The United States was now fighting Spain in Asia, and the Philippines would become the site of the war's first major naval battle.

The stakes for Dewey and his sailors were extremely high.

His Asiatic Squadron was thousands of miles from any other American naval force. That meant no reinforcements, no backup, and no resupply. If something went wrong—if their ships were damaged or sunk—there would be no rescue. The squadron was completely on its own.

And worse, if Dewey's ships became crippled in battle, there were no safe harbors nearby where they could escape and make repairs. The wounded ships would then become easy targets for the Spanish to finish off, putting Dewey's men at grave risk.

Dewey believed the Spanish fleet was probably anchored at a deep-water naval base within Manila Bay, known as Subic Bay Naval Base—a strong and protected position for the Spanish.

But when he arrived, no Spanish warships were there.

He soon learned that Admiral Montojo had moved his fleet to Cavite Naval Base, a shallower harbor also located within Manila Bay.

Although that location was not as defensible as Subic Bay and left the Spanish fleet more vulnerable, Montojo had little choice. Many of his ships needed repair and could not sail far.

Now, with the enemy located and the element of surprise still in his favor, Dewey prepared for battle.

Spanish Admiral Montojo.

Why did Admiral Montojo choose the shallows?

Sensing that a sea battle was coming, Admiral Montojo believed that the shallow waters of Manila Bay would give his damaged ships a better chance of survival. By staying close to land, the Spanish fleet could receive support from coastal artillery batteries stationed onshore. Though his fleet outnumbered Dewey's squadron, Montojo knew his outdated and poorly equipped ships were no match for the modern, steel-hulled American warships.

To his credit, Montojo also made this decision with the survival of his crews in mind. Some parts of the water near Cavite Naval Base were so shallow that if one of his ships were sunk, the superstructure—the upper parts of the ship—would still remain above water. That meant sailors abandoning a sinking vessel might survive by clinging to the wreckage, rather than being dragged down into deep ocean waters.

But before Dewey could engage the Spanish fleet, he faced a critical decision: how to enter Manila Bay.

There were two main channels.

One was called *Boca Chica* (Spanish for "Little Mouth"). This passage was only about two miles wide, which meant American ships passing through it would be vulnerable to direct fire from Spanish shore batteries.

The other option was *Boca Grande* ("Big Mouth"), about four miles wide. But it came with serious risks. Spanish forces had lined the channel with explosive sea mines, and its shores were defended by artillery as well. To make matters worse, the U.S. had no accurate navigation charts for this route. There were real dangers of mines, hidden shoals (shallow places), or running aground.

Even so, Dewey decided to take the Boca Grande channel. The wider passage was better suited for maneuvering his larger ships. But his squadron would have to move slowly and carefully, watching for mines and enemy fire as they followed the curving coastline in darkness.

On April 30, Dewey's squadron entered the Boca Grande channel, headed for Cavite. Dewey, aboard his flagship USS *Olympia*, led the way. He believed that if a mine was triggered, his ship should take the first hit—a powerful display of his leadership and courage.

The next morning—May 1, 1898—Dewey's ships were in position and closing in on the Spanish fleet anchored near Cavite.

At 5:41 a.m., Dewey was waiting for daylight to clearly see his targets before firing. But suddenly, the Spanish opened fire first—BOOM!

USS *Olympia* in the left foreground, leading the U.S. Asiatic Squadron in destroying the Spanish fleet off Cavite. A vignette portrait of Rear Admiral George Dewey is featured in the lower left.

Fortunately for the Americans, the Spanish gunners missed badly. Their shells splashed into the bay.

Dewey's squadron continued closing in. When his ships were about 5,400 yards (a little over three miles) from the Spanish line, Dewey turned to the captain of the *Olympia*, Charles V. Gridley, and famously said:

"You may fire when ready, Gridley."

That simple command has become one of the most iconic quotes in U.S. naval history. It revealed Dewey's calm, confident manner pressure—a true hallmark of leadership in battle.

Dewey's squadron formed a battle line, with each ship sailing behind the next in a straight column. Then *Olympia*'s massive turret guns opened fire—two turrets, each armed with two 8-inch guns. They roared to life.

Dewey's squadron, left, in a battle line firing on the Spanish ships that anchored close to shore.

The *Olympia*'s turret guns had the advantage of being able to rotate from port to starboard—left to right—allowing the gunners to adjust their aim as they unleashed shell after shell at the Spanish fleet.

But just as the guns were thundering, a new threat appeared.

A Spanish torpedo boat was spotted—charging straight at the *Olympia*!

Torpedo boats are small, fast, and designed to sneak in close and launch deadly torpedoes at larger ships such as battleships and cruisers. If a torpedo boat gets too close, it can inflict serious damage—possibly even sink a flagship.

The American gun crew didn't hesitate.

They quickly swiveled the guns, took aim, and blasted the torpedo boat out of the water with a devastating shot—neutralizing the threat before it could strike.

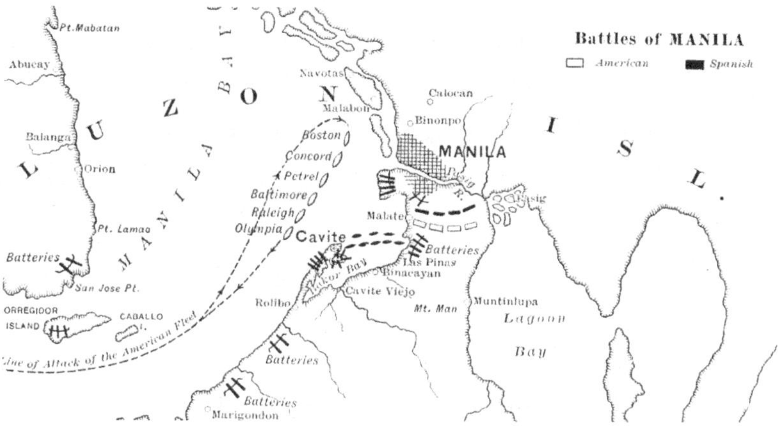

Copy of a map showing the loop (center) that Dewey's squadron made in Manila Bay, while the Spanish fleet clung to the shores of Cavite.

Moving down the port (left) side of the Spanish fleet, the American cruisers opened fire, targeting Admiral Montojo's flagship, the *Reina Cristina.*

Dewey's gun crews hammered the ship with shell after shell, striking from stem to stern—front to back—ripping through decks, igniting fires, and exploding on impact.

Realizing the catastrophic damage, Montojo quickly abandoned the *Reina Cristina* and transferred to another ship, the *Isla de Cuba*, in a desperate attempt to continue commanding the battle.

Within minutes, *Reina Cristina* was engulfed in flames. Eighty sailors were killed, including the ship's captain. The flagship was completely destroyed.

At this point, Dewey's squadron had reached the end of its firing run along the Spanish line.

But warships can't exactly throw it in reverse. So instead, Dewey ordered his entire line to perform a turnaround maneuver: a continuous loop, bringing each ship back around to begin another pass, all while still firing on the enemy fleet.

As the maneuver was underway, the USS *Raleigh* landed a devastating hit on the Spanish cruiser *Castilla*. The shell smashed the ship's bridge—its control tower—and disabled the vessel's guns. *Castilla* began sinking slowly, fires spreading along her deck.

Despite these crushing blows, the Spanish were not giving up.

The cruiser *Don Juan de Austria* suddenly charged forward, attempting a bold attack on Dewey's battle line.

But *Olympia*'s gunners responded instantly. They unleashed a furious barrage, raking the advancing cruiser with such force that the *Don Juan de Austria* was forced to turn away, battered and smoking, just to survive.

One of the wrecked Spanish ships in the Battle of Manila Bay.

Now, the only defense remaining for the Spanish was their shore batteries—the cannons mounted along the coastline. But those guns couldn't reach Dewey's ships, which had pulled back just beyond their range.

Dewey's gunners, however, were not out of range. They returned fire with precision, neutralizing the shore batteries and silencing the last line of Spanish defense.

By 12:30 p.m., the battle was effectively over.

Nearly every Spanish ship had been sunk, beached, or captured. Among the most notable losses were the *Reina Cristina, Castilla, Don Antonio de Ulloa,* and several others.

The Spanish fleet at Manila Bay had been completely destroyed.

The toll on the Spanish was devastating.

More than 160 sailors were killed in action. Another 214 were wounded.

In contrast, the American losses were astonishingly light. Only eight U.S. sailors were wounded. And just one died in action—and that was due to heat exhaustion, not enemy fire.

Commodore Dewey later recorded the event in his journal:

"Sunday, May 1st. Dewey reached Manila at daylight and immediately engaged the Spanish ships and batteries at Cavite. Destroyed eight of the former, including *Reina Cristina* and *Castilla*. Anchored at noon off Manila ..."

Talk about a cool-headed commander under fire!

The Battle of Manila Bay was a crushing blow to the Spanish navy and marked the beginning of the end for Spain's colonial empire.

The victory had global repercussions.

Just weeks later, Spain agreed to peace talks.

The result?

A treaty that handed over several Spanish colonies to the United States, including Cuba. Also given to the U.S. were Puerto Rico, the Philippines, and the island of Guam in the Pacific Ocean.

Commodore Dewey's victory at Manila Bay didn't just win a battle. It launched the United States onto the world stage as a major naval and imperial power.

WHO FOUGHT?

George Dewey was born on December 26, 1837, in Montpelier, Vermont.

Even though he grew up far from the ocean, George loved adventure and excitement. As a boy, he was known for being tough and brave—he once stood up to the town bully with nothing but his fists! His courage made him a leader among his friends.

When he turned 17, George decided he wanted to be a naval officer. He attended the U.S. Naval Academy and worked hard to learn everything about ships and sailing. After graduating, he served in the Navy during the Civil War, fighting in large battles that revealed his fearless spirit.

But the moment that made Dewey famous came many years later, halfway around the world: the Battle of Manila Bay in the Philippines.

Now a commodore, he quietly sailed his squadron into Manila Bay under cover of darkness. As the sun rose on May 1, Dewey gave his famous command: "You may fire when you are ready, Gridley." By noon, every Spanish ship was neutralized or destroyed—yet not a single American ship was sunk. It was a stunning victory for the still-young United States against a large naval power such as Spain.

When news of Dewey's victory reached the United States, people cheered in the streets. "Remember the *Maine*!" they shouted.

Dewey became a national hero. Parades were held in his honor. Streets, schools, and even babies were named after him. Dewey had shown bravery, smart thinking, and strong leadership—qualities every hero needs—while the underdog in battle.

Later, Dewey was promoted to Admiral of the Navy, the highest rank ever given to a U.S. naval officer. In fact, Dewey is the only American to achieve that rank.

He served his country for many more years, always proud of his adventures at sea. Later in life, Dewey said, "The best way to win a war is to be prepared for it."

George Dewey died in 1917 at the age of 79.

BOOKS

Manila Bay 1898: Dawn of an American Empire by Brian Lane Herder. Illustrated with stunning battle scenes, photographs, and informative maps and diagrams, this book covers the May 1 naval battle and others.

INTERNET

An eyewitness account of the battle, with images: https://youtu.be/kEaUGiiw2Tg

The battle further explained: https://youtu.be/QynrA8l39eM

The Battle of Tsushima

The Russo-Japanese War

May 27–28, 1905

Japanese warships, on the left, firing on the Russian fleet, on the right.

THIS NEXT EPIC SEA BATTLE has been described as "by far the greatest and most important naval event since Trafalgar."

(And hey—isn't it cool that you know about Trafalgar? For a fun challenge, ask a few adults if they know what happened at Trafalgar. If they don't, go ahead and impress them with your knowledge!)

In the early 1900s, both Russia and Japan were looking to expand their empires. These two rising world powers were competing for influence and control in a region often called "the Far East"—including areas such as China, Korea, Vietnam, and Thailand.

Their rivalry eventually exploded into open warfare.

On the night of February 8, 1904, without a formal declaration of war, a Japanese naval squadron launched a surprise attack on the Russian fleet anchored at Port Arthur, a strategic harbor in Manchuria (northeastern China).

The Japanese torpedo boats struck suddenly, damaging one Russian cruiser and two battleships before slipping away into the night.

That attack marked the start of the Russo-Japanese War—a conflict that would dramatically shift the balance of power in Asia.

Following the attack, Japan formally declared war, and its navy launched several more successful assaults on Russian forces in the Pacific. The Japanese navy had the early upper hand, thanks to superior training, modern ships, and bold leadership.

Russia, meanwhile, was stunned and unprepared.

In response, Russian commanders made a drastic decision: they would send their main naval fleet, located in Europe, all the way to Asia to reinforce their struggling Pacific squadron.

The plan was for all Russian naval forces to rendezvous at the friendly port of Vladivostok, located in eastern Russia near the Sea of Japan.

But to get there, the Russian fleet would have to sail over 18,000 miles, around Africa's Cape of Good Hope, across the Indian Ocean, and through dangerous enemy waters.

What happened next would become one of the most dramatic—and most disastrous—long-distance naval journeys in history.

Russian Admiral Rozhestvensky, commander of the Baltic fleet.

However, the Russian reinforcements needed to sail all the way from the Baltic Sea in northern Europe to the North Pacific—a massive journey—to link up with the rest of the fleet near Japan.

Under normal circumstances, this trip wouldn't have been so daunting, thanks to a vital shortcut: the Suez Canal in Egypt.

The Suez Canal, completed in 1869, forms a crucial passageway between the Mediterranean Sea and the Red Sea, allowing ships to bypass the long and dangerous route around Africa's Cape of Good Hope.

The canal dramatically shortens the journey from Europe to the Middle East and to the Pacific Ocean.

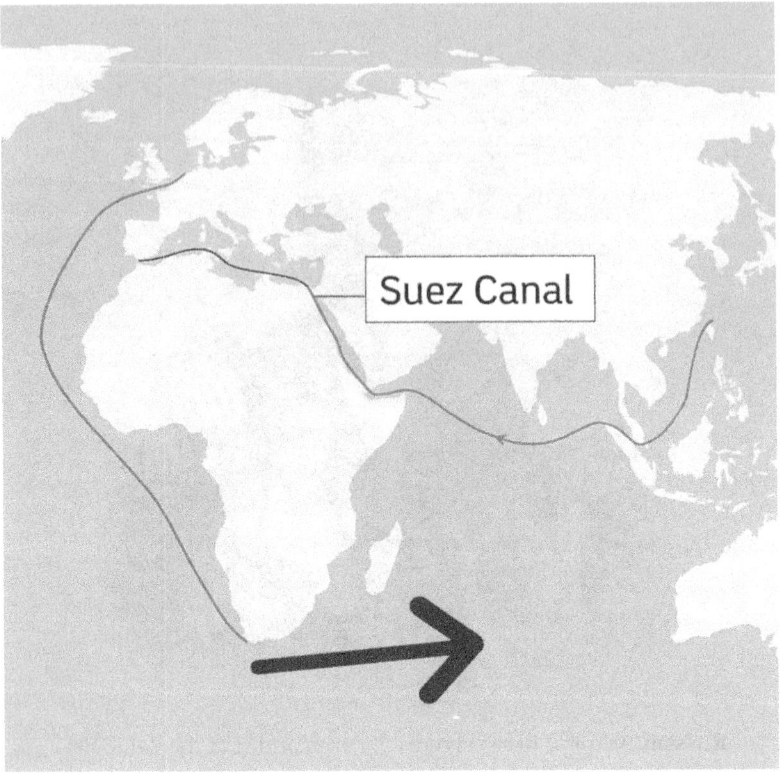

Sea routes from the Mediterranean Sea (above the Suez Canal) and the Atlantic Ocean (on the left). Notice the shortcut that the canal provides, allowing ships to move down Africa's east side instead of sailing all the way around Africa.

But complications soon arose—and not the kind anyone expected.

About six months after Japan's surprise attack on Port Arthur, disaster struck. A Russian squadron sailing through the North Sea—off the coast of Britain—mistook a group of British fishing trawlers for Japanese warships.

Yes, really.

In a panic, the Russian fleet opened fire on the British fishing boats. One trawler was sunk, several others were damaged, and several British fishermen were killed or injured.

Naturally, Britain was outraged.

How could the Russian navy possibly confuse a few slow-moving fishing vessels for Japanese warships—especially considering that Japan was thousands of miles away and had no naval bases anywhere near the North Sea?

The incident, known as the "Dogger Bank Incident," nearly brought Britain and Russia to the brink of war. Only last-minute diplomacy prevented a wider conflict.

This embarrassing mistake damaged Russia's reputation on the world stage and added yet another layer of tension, confusion, and bad luck to the Russian fleet's already difficult journey.

British cartoon showing an Englishman shaking hands with a Japanese soldier, each backed by their country's flag.

The incident nearly prompted Britain to enter the Russo-Japanese War. At the time, Britain had already formed an alliance with Japan through the Anglo-Japanese Alliance of 1902, which promised mutual support in protecting their interests in Asia.

In the end, Britain agreed to remain officially "neutral" in the conflict—but that neutrality came with a wink.

Though it didn't send troops or warships, Britain quietly supported Japan in other important ways.

For example, when the Russian fleet attempted to sail from the Baltic Sea to reinforce its forces in the Pacific, it planned to use the Suez Canal—a vital shortcut connecting the Mediterranean Sea to the Red Sea and Indian Ocean.

But the British government denied the Russian fleet passage through the canal.

Why? Because Britain controlled the Suez Canal—and had since the late 1800s. The British had gained influence by purchasing a major stake in the Suez Canal Company from Egypt and then solidified that control with a strong military presence to protect its colonial trade routes.

By denying access, Britain forced the Russian fleet to take the much longer and more dangerous route around Africa's Cape of Good Hope, adding thousands of miles and many months to their journey—all while their ships and sailors grew more exhausted and less prepared for battle.

Russian ironclad warship *Osliabia*. It sailed the long route. Its crew stands on deck.

Instead of sailing through the Suez Canal, the Russian warships were now forced to take the long way around the continent of Africa—a journey of roughly 18,000 nautical miles.

(To help you imagine that distance: on land, it's like driving across the United States from coast to coast *seven* times!)

This massive voyage was filled with challenges and dangers. Along the way, Russian sailors fell ill from various tropical diseases, including malaria, dysentery, and other infections that thrived in the hot, humid climates. The ships were also constantly low on coal—the fuel that powered their engines and kept the fleet moving.

Making things worse, many ports along the way belonged to neutral nations—countries that had officially declared they would not take sides in the Russo-Japanese War. As a result,

Russian warships were not allowed to dock and refuel at most of these ports.

That meant supply ships had to carry coal and attempt to make mid-sea transfers, a difficult and dangerous task that was often delayed or interrupted. Sometimes, the Russian sailors were trapped aboard their ships for weeks, unable to go ashore or even resupply.

Morale plummeted. Sailors grew exhausted, sick, and frustrated. And with every passing mile, the fleet was becoming less battle-ready.

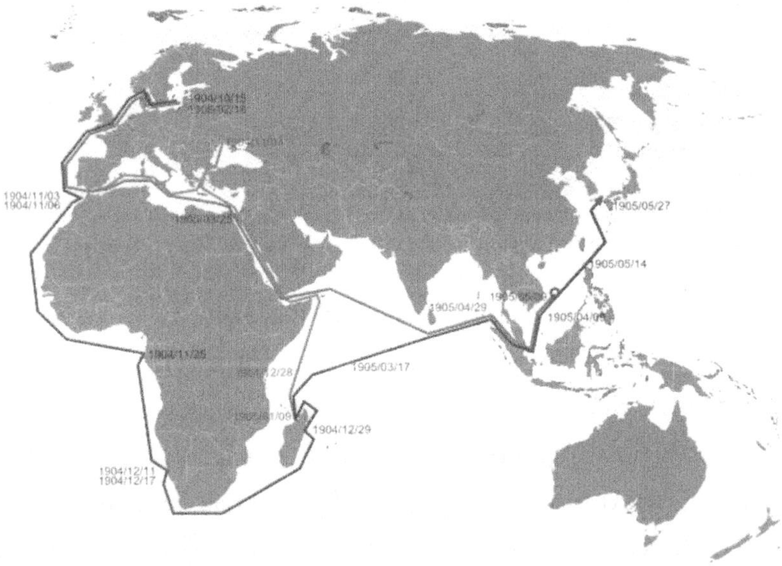

Another map showing the two sailing routes from Europe (upper left) to the North Pacific (far right). The line traced around Africa is the sailing route of the Russian ships denied use of the Suez Canal—that straight line running downward from the Mediterranean Sea along the east coast of Africa.

Finally, after six grueling months at sea, the Russian fleet emerged into the Pacific Ocean.

But by then, the Japanese had figured out the plan.

It was clear that Russia intended to link up with its remaining fleet at the friendly naval base in Vladivostok, on Russia's eastern coast. The shortest sea route to get there was through the Tsushima Strait—a narrow passage between Korea and Japan.

Admiral Tōgō Heihachirō, the brilliant commander of the Imperial Japanese Navy, correctly predicted that the Russians would come through that strait.

He set his fleet into motion.

Japanese warships began scouting the waters around the Tsushima Strait, maintaining a silent watch—waiting for the enemy to appear.

The trap was set.

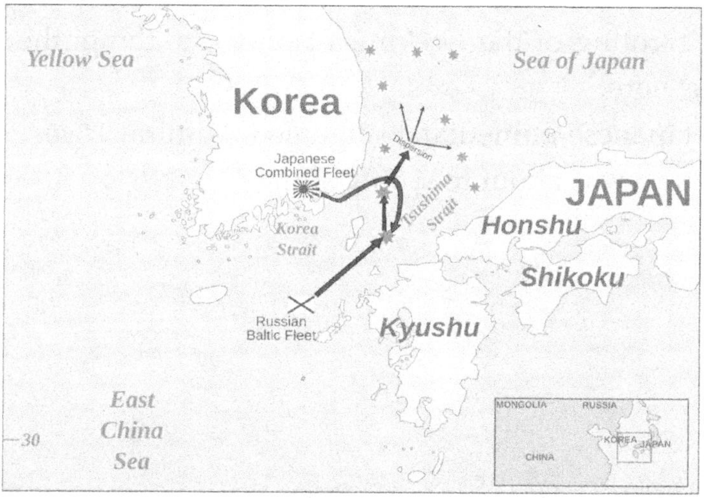

Tsushima Strait (arrows) is located between Korea's southern tip and Japan's southern islands. Vladivostok is located further north, on the coast of Russia (which is shown in the box, lower right).

Realizing they were now in hostile and treacherous waters, the Russian fleet tried to sneak through the Tsushima Strait without being detected.

To avoid discovery, the ships veered away from normal shipping lanes, maintained strict radio silence, and sailed at night with all lights turned off.

On the night of May 26, 1905, cloaked in darkness and thick fog, the Russian fleet quietly approached the Tsushima Strait—their final stretch before reaching Vladivostok.

The fog was a lucky break. It helped to conceal the darkened ships as they moved slowly through the strait.

But just before 3 a.m., a sharp-eyed crew aboard a Japanese cruiser spotted three faint lights flickering on the horizon.

The Japanese ship moved in to investigate.

As they drew closer, they identified one of the vessels: it was the *Orel*, a Russian hospital ship—a non-combat ship that, under international law, was required to stay illuminated even during wartime to signify its humanitarian role.

The sighting of the *Orel* meant only one thing: the Russian fleet was here.

The Japanese immediately signaled Admiral Tōgō.

The trap was about to spring shut.

Hospital ship *Orel*. During daylight hours, red crosses on its smokestacks signaled its non-combatant status.

According to international law, hospital ships must keep their lights on at night to signal that they are non-combatants.

These ships are protected under wartime agreements because they carry the wounded, medical staff, and supplies—and they are not armed for battle.

Because of this, the Japanese crew—once they identified the *Orel* as a hospital ship—might have allowed it to pass without incident.

But then, something unexpected happened.

The crew aboard the *Orel* mistook the Japanese cruiser for one of their own ships in the Russian fleet. Without realizing they were signaling the enemy, they flashed a message in Russian naval code.

That was all the Japanese needed to confirm it: this wasn't just a lone hospital ship—the Russian fleet was here.

The location was compromised. The element of surprise was gone.

And now, Admiral Tōgō knew exactly where to strike.

Admiral Tōgō Heihachirō (center) on the bridge of the battleship *Mikasa* with his officers. To the left, the signal flag for battle is being raised.

The Japanese cruiser immediately sounded the alarm—using a wireless telegraph, one of the most advanced communication tools of the time.

Morse code was still relatively new in naval warfare, and this event marked the first time a wireless telegraph was used in an operational sea battle.

The Japanese had wisely invested in developing their own wireless systems. Their equipment was built, maintained, and repaired by specially trained navy technicians. That meant if something broke, it could be fixed at sea—fast.

The Russians, on the other hand, relied on outdated German wireless sets. And unfortunately for them, no one on board had the expertise to repair them if they failed.

As soon as Admiral Tōgō Heihachirō received the signal from his scouting cruiser, he wired an official message to the navy minister in Tokyo:

"In response to the report that enemy ships have been sighted, the Combined Fleet will immediately commence action and attempt to attack and destroy them. Weather today fine but high waves."

The stage was set.

Japan was ready.

And the greatest sea battle of the 20th century was about to begin.

Early on May 27, the Japanese fleet proceeding toward the enemy.

The Japanese fleet was well prepared and highly organized. It included:

- 4 battleships
- 8 cruisers
- 21 destroyers
- Plus a mix of torpedo boats and support ships.

At the heart of the fleet was Admiral Tōgō's flagship, the *Mikasa*—a state-of-the-art battleship. It was equipped with four turret-mounted 12-inch guns, protected by thick steel armor, and outfitted with a wireless telegraph system for sending Morse code—a huge advantage in coordinating the fleet during battle.

The Russian fleet, commanded by Admiral Zinovy Rozhestvensky, was even larger and seemingly more powerful. Rozhestvensky led from his flagship, the *Knyaz Suvorov*.

The rest of his force included:

- Eight battleships
- Eight cruisers

- Nine destroyers
- Multiple support vessels, including the hospital ship *Orel*.

In fact, the Russian fleet was so big and heavily armed that many observers believed Japan stood no chance. On paper, the Russians had the edge in firepower and size.

But reality told a different story.

After their 18,000-mile voyage, Russian sailors were exhausted and sick, and morale was dangerously low. Their ships had also become "fouled"—meaning seaweed, kelp, and barnacles had attached to their hulls during the long journey. This buildup of marine organisms increased drag, slowing the ships and reducing their maneuverability.

And in a sea battle, speed is everything.

Some estimates suggest the Japanese ships were 5–10 knots faster (about 6–11 mph) than their Russian counterparts. Because a fleet must move at the speed of its slowest ship, this gave the Japanese a significant tactical advantage.

By early afternoon on May 27, 1905, the two fleets finally came into range.

The Russian vessels were sailing in a single-file line, preparing for combat.

That's when Admiral Tōgō made a bold and brilliant move: he ordered a maneuver called "crossing the T."

Here's how it worked:

The Russian ships were lined up in a straight formation, one behind the other. But they did not have rotating turrets—their main guns could only fire forward or to the sides with limited range. Firing too far ahead risked hitting their own ships in the line.

Tōgō, meanwhile, turned his ships at a right angle to the Russian line, sailing across the front of their formation. This T-shaped position allowed every Japanese ship to fire full broadsides—all guns on one side of each ship—into the front of the Russian line.

The Russians, by contrast, could only fire a few forward guns—if any at all.

It was a devastating tactical advantage, and it would turn the tide of the battle almost immediately.

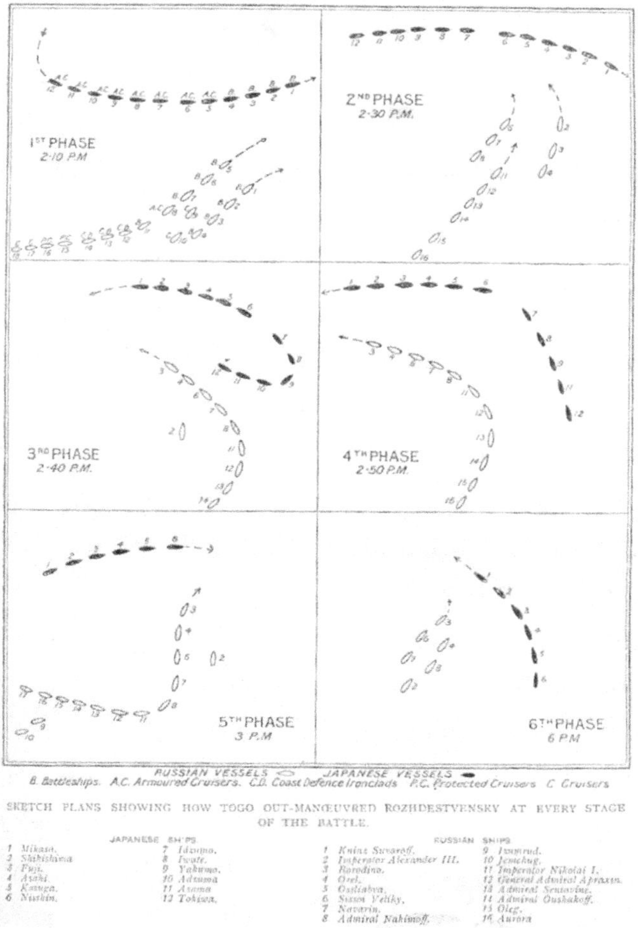

Copy of a sketch showing the six maneuvers used by the Japanese ships (darker ovals) to destroy the Russian fleet (lighter ovals), repeatedly "crossing the T."

BOOM!

The Japanese guns unleashed a devastating firestorm of high-explosive shells—a storm unlike anything the Russian navy had ever faced.

These were not ordinary shells. They were filled with pure picric acid, an incredibly powerful explosive. Picric acid is highly unstable—especially when it comes into contact with iron or other heavy metals, which can trigger accidental detonations.

But a clever Japanese engineer had solved that problem.

He developed a technique to coat the inside of each shell with lacquer, then seal it with wax, preventing the acid from reacting with the metal casing. On top of that, the Japanese added a contact fuse—a mechanism that caused the shell to explode the instant it struck a target.

The result? An explosion that burned hotter and more violently than anything the Russians had prepared for.

To add to their technological edge, the Japanese also used smokeless propellant in their naval guns. That meant their shots didn't produce the giant clouds of white smoke that black powder did—so Japanese spotters could see their targets clearly, adjust their aim, and keep firing with deadly accuracy.

Their first main target: the Russian flagship, the *Knyaz Suvorov*, with Admiral Rozhestvensky on board.

Shells began to rain down with terrifying precision.

Later, the *Suvorov*'s captain recalled:

"It seemed impossible even to count the number of projectiles striking us ... I had not only never witnessed such a fire before, but I had never imagined anything like it. Shells seemed to be pouring upon us incessantly, one after another ..."

The Japanese high-explosive shells tore through the *Suvorov*'s upper decks, shredding steel plating and destroying the superstructure. Metal splinters flew like shrapnel, wounding and killing crew members. Iron ladders melted and crumpled like paper. Entire gun turrets were blown off their mountings.

The captain continued:

"In addition to this, there was the unusually high temperature and liquid flame of the explosion, which seemed to spread over everything. I actually watched a steel plate catch fire from a burst."

(Note: Steel doesn't literally "catch fire" like wood, but it loses all structural integrity when heated to around 1,000 degrees Fahrenheit. At those temperatures, it bends, warps, and collapses.)

The Russians should have been able to return fire with similar intensity. Years before the war, a brilliant Russian artillery officer had proposed a new type of 12-inch gun and advanced shells that could have rivaled the Japanese design.

But during early testing, that officer was killed in a tragic accident, and his innovations were never adopted.

Now, in this crucial battle, the Russian fleet was stuck using older armor-piercing rounds, less sensitive fuses, and black powder—which produced thick smoke clouds that blinded their own gunners with every shot.

The Japanese had combined superior strategy, morale, and technology into one devastating force.

And the Russian fleet was burning beneath it.

Battle of Tsushima, 1905, painting by Tōjō Shōtarō, showing the explosive shells creating rough seas.

The Japanese shells were so powerful that even when they missed their targets, the force of the explosions created massive splashes. Those water blasts sent huge waves crashing into nearby Russian ships, rocking them violently and throwing them off balance—interfering with aim, speed, and stability.

One of the most tragic moments came aboard the Russian battleship *Osliabia*.

Japanese shells tore into the ship's upper works, severing thick steel cables that connected to its fire control systems—essential for coordinating the ship's guns. Other shells struck the main masts, damaging communications and knocking out forward gun turrets.

Metal splinters the size of swords slammed into the conning tower—the ship's control center—killing and wounding officers inside. The confined steel structure turned into a deadly oven, filled with razor-sharp shrapnel.

But the most devastating blow struck the *Osliabia*'s waterline—the critical point where the ship's hull meets the ocean.

One direct hit ripped open a massive hole, and seawater surged in uncontrollably.

Despite efforts to save the ship, the damage was too great. The *Osliabia* sank beneath the waves, taking with her 470 crewmen.

This was more than just a loss—it marked a milestone in naval history.

The *Osliabia* became the first all-steel battleship ever sunk purely by naval gunfire—not by torpedoes, not by aircraft (which wouldn't be used in naval warfare for another decade), and not by shore-based artillery.

This was ship vs. ship, and the Japanese firepower had redefined what naval combat could do.

Russian cruiser *Oleg* in 1905, after the Battle of Tsushima, with holes large enough for men to stand in.

The Japanese shells didn't just hit hard—they hit lethally.

Explosions ignited the Russian ships' gunpowder bags, setting off devastating internal blasts that blew entire turrets into the air like tin cans.

One Russian vessel was struck 34 times in a row!

As fires broke out on board, thick black smoke slithered through interior compartments, blinding the Russian sailors and choking their lungs. Panic and confusion spread across the fleet.

At one point during the chaos, the *Knyaz Suvorov*—Admiral Rozhestvensky's flagship—was torpedoed. A blast of flying shrapnel struck Rozhestvensky in the head. Severely wounded, he was evacuated to another ship just before the *Suvorov* slipped beneath the waves.

The Japanese navy's relentless attack continued until the daylight faded. As both fleets tried to regroup, there was a brief lull—but it didn't last long.

At sunset, the Japanese launched a second wave of devastation: 37 torpedo boats and 21 destroyers. All aimed straight into the broken Russian formation.

The attack continued for three more hours, through the darkness of night. Another Russian battleship was sunk, and a second was crippled. Two armored cruisers, too damaged to escape, were scuttled—intentionally sunk by their own crews as water flooded through the holes in their hulls.

Hundreds of Russian sailors lay dead in the water.

By dawn, only six Russian ships remained. Under cover of darkness, they attempted to escape. But the Japanese hunted them down.

At 10:34 a.m. on May 28, 1905, the surviving Russian vessels hoisted the flag of surrender.

But Admiral Tōgō, cautious to the end, feared it might be a trick.

He ordered his guns to open fire again—until a clear and confirmed surrender was received minutes later.

The Battle of Tsushima was over.

It was a staggering victory for Japan.

The Japanese had sunk or captured two-thirds of the Russian fleet. More than 5,000 Russian sailors were killed and 5,600 were taken as prisoners of war. Only about 60 officers and 1,000 crewmen escaped to Vladivostok.

Meanwhile, Japan had lost only 3 torpedo boats and 117 men. None were captured.

It was an utterly humiliating defeat for Russia. The empire had lost all of its battleships and most of its cruisers and destroyers.

Even more importantly, the battle signaled the end of the Russo-Japanese War—with Japan as the clear victor.

For the first time in modern history, an Asian power had defeated a European empire in a major war.

How did this underdog force achieve such a decisive victory?

- Superior tactics, such as "crossing the T"
- Faster ships with better maneuverability
- More accurate and advanced artillery
- And perhaps most crucially: higher morale and sharper discipline among the Japanese crew.

In August 1905, the two sides met in Portsmouth, New Hampshire, to negotiate peace. U.S. President Theodore "Teddy" Roosevelt served as the mediator.

The resulting Treaty of Portsmouth ended the war—and earned Roosevelt the Nobel Peace Prize.

U.S. President Teddy Roosevelt, center, at the signing of the Treaty of Portsmouth, ending the Russo-Japanese War. Russian negotiators on the left, Japanese on the right.

Japan was declared the victor of the Russo-Japanese War. But the political aftershocks of the Battle of Tsushima would continue to echo for decades.

Some historians argue that Tsushima made Japan overconfident. This stunning victory convinced many Japanese military leaders that their navy could defeat any Western power—including the United States.

That mindset of invincibility and belief in Japan's naval superiority grew over the next several decades.

Eventually, that overconfidence contributed to one of the most pivotal events of the 20th century: the surprise attack on Pearl Harbor, Hawaii, on December 7, 1941—a strike that would draw the United States into World War II.

In many ways, the seeds of that global conflict were planted in the waters of the Tsushima Strait, where Japan first proved to the world that an Asian nation could challenge and defeat a European empire—but where a belief in unstoppable dominance also began to take root.

WHO FOUGHT?

Japanese Admiral Tōgō Heihachirō in full military dress.

In 1848, a boy named Tōgō Heihachirō was born on the small island of Kagoshima, Japan. No one knew it yet, but this boy would grow up to become one of the world's greatest naval commanders and a hero to his country.

During Tōgō's youth, Japan was just starting to open up to the Western world. At age 17, he joined the Japanese navy and learned about ships and cannons. The Japanese government later sent him to England to study naval science and military strategy, focusing especially on navigation, artillery, ship handling, and engineering.

Tōgō spent several years serving aboard British warships as a cadet, learning practical seamanship, navigation, and artillery use. That experience allowed him to observe the latest European naval tactics, which were more advanced than Japan's at the time.

He also spent time studying the English language and culture so that he could understand English orders and read their manuals.

It was this firsthand experience with modern naval warfare that helped Tōgō later in developing Japan's navy into a powerful military force. Because he had learned the latest European naval tactics, like gunnery drills, ship formations, and maneuvering strategies, he could plan that "crossing the T" maneuver at Tsushima—a tactic that wasn't commonly used in Japan.

Tōgō also admired the calm, disciplined leadership of the British naval officers, and he adopted a similar style by staying composed even under pressure.

In the Battle of Tsushima, that calmness inspired his sailors to fight with extreme courage despite facing a large enemy.

In short, Tōgō's British education blended Western ideas with Japanese determination. That's why people say his victory at Tsushima wasn't just a military win—it showed how Japan had mastered modern warfare and strategy.

Tōgō became a hero in Japan and stayed respected for the rest of his life. He died in 1934, but his name is still remembered as a symbol of courage, wisdom, and daring leadership.

Some even called Tōgō the "Nelson of the East," comparing him with Britain's famous Admiral Nelson.

BOOKS

The Russo-Japanese War 1904–1905 by Geoffrey Jukes. Published by Osprey, this book has many good illustrations and photos.

The Russo-Japanese War 1904–05 (Men-at-Arms series) by Alexei Ivanov and Philip Jowett. Heavily illustrated, with rare photos and excellent atmospheric paintings by Russia's leading military illustrator.

Russian Soldier vs Japanese Soldier: Manchuria 1904–05 by David Campbell. This fully illustrated study tells the story of the fighting men of the Russo-Japanese War and how these two empires clashed, heralding a new phase in modern warfare as World War I loomed on the horizon.

German Naval Tricks of World War II

The *Kormoran*, the largest of Germany's converted Sea Raider merchant ships during World War II.

DURING WORLD WAR II, Germany was reluctant to engage directly with the British Royal Navy. The Royal Navy was one of the most powerful fleets in the world, with larger numbers, more advanced ships, and experienced sailors trained for high-seas combat.

Instead of meeting them head-on, Germany developed ingenious naval weapons and tactics designed to disrupt, deceive, and destroy from the shadows.

Two of the most inventive were the Sea Raiders and the infamous U-boats—German submarines that haunted Allied shipping routes.

SEA RAIDERS

If you've ever played with those toys called Transformers, you know how quickly they can change from a robot into a car, a tank—or even a bumblebee.

Well, during World War II, Germany's own version of Transformers took to the seas. They were called Sea Raiders—and they were devils in disguise.

At first glance, these ships looked like ordinary cargo vessels or merchant freighters. They were made to blend in on the open ocean. Some even had fake smokestacks, complete with crew members pretending to stoke a fire—just to sell the illusion.

But it was all a trick.

The smokestacks, water towers, and even some of the deck housing—like cabins or storage sheds—were actually camouflaged weapons platforms. Behind those innocent-looking structures were hidden guns, torpedo tubes, and anti-aircraft weapons, ready to launch devastating surprise attacks.

These ships could transform—in a matter of seconds—from peaceful-looking freighters into fully armed raiders, capable of sinking Allied merchant ships before the enemy even realized what was happening.

The Sea Raiders were designed for deception and hit-and-run tactics—slipping through shipping lanes undetected, then striking with lethal force.

Early "sea raiders" from World War I. The sailors surround a 15 cm (5.9-inch) gun on the deck of the German armed merchant raider SMS *Wolf*. When not in use, the gun was hidden under sheets of canvas. The ship's railing could also be lowered to allow the gun barrel to protrude from the ship. This early "raider" could be easily disguised as a merchant ship.

Germany converted nine merchant ships into these fearsome Sea Raiders, which came in two sizes: light cruisers and heavy cruisers. Though they looked like innocent cargo vessels, they were anything but.

Each Sea Raider carried a crew of about 400 men, trained to operate like modern-day pirates. These crews were heavily armed and ready to fight. A typical Sea Raider bristled with:

- Six 150 mm (5.9-inch) naval guns
- Two 37 mm (1.5-inch) anti-aircraft guns
- Five 30 mm (1.2-inch) machine guns
- Six 21-inch torpedo tubes
- And even two seaplanes for reconnaissance and scouting missions.

These weren't cargo ships—they were floating war machines in disguise.

One of the Sea Raiders' most clever features was a hinged bulkhead, engineered to hide their main guns. When it was time to attack, the steel panel would drop away, revealing the ship's full firepower in an instant. Imagine sailing up beside what you thought was a harmless merchant freighter—only to see its hull open like a trapdoor and unleash a storm of shells and torpedoes.

Even inside, these ships were full of secrets, with hidden compartments built into the hull. Some were used to hold captured prisoners from Allied ships. Others stored foldable seaplanes, which the Germans could launch with catapults to scout enemy convoys, monitor coastlines, or guide attacks.

The Sea Raiders were designed for long missions. They often operated alone in remote parts of the ocean for months at a time, refueling at sea or using disguised supply ships to stay active. They targeted Allied merchant shipping, aiming to cut off supplies of food, fuel, and weapons headed to Britain and its allies.

Though not as famous as the U-boats, the Sea Raiders were masters of stealth and surprise, striking fear into convoys that never saw them coming—until it was too late.

German cruiser *Atlantis*, a converted German Sea Raider. It traveled more than 100,000 miles in 602 days and sank or captured 22 ships.

The crews of Germany's Sea Raiders weren't just well armed—they were cunning and deceptive, using tricks and psychological warfare to fool and disable their enemies.

For example, their scout seaplanes often flew with fake insignias painted on their wings and fuselages—symbols meant to mimic those of British or neutral aircraft. This clever disguise allowed German pilots to fly low over Allied ships without raising suspicion.

Once in range, they would use one of their most devious tactics: hooking the ship's radio antenna mast.

Flying just above the deck, a German scout plane could drop a weighted hook that would snag the rigging of a ship's antenna and rip it clean off—cutting off all radio communication. Without an antenna, the Allied ship couldn't send out an SOS distress signal—leaving it isolated and helpless.

The Sea Raiders themselves also used false flag tactics, flying the flags of friendly or neutral nations like Sweden, Norway, or even the United States. They'd approach these unsuspecting cargo vessels under such disguises, giving no indication of danger.

But behind these disguises, the Germans had already used rangefinders to plot an ambush.

Then came the moment of betrayal.

The false flags were suddenly lowered, and in one swift motion, deck houses collapsed, the fake water towers dropped away, and massive naval guns emerged from their hiding places. As the German war flag was hoisted, the attack began.

During their three years of operation, the Sea Raiders managed to sink or capture ships carrying more than 886,000 tons of cargo. These supplies—meant to support the war effort of the Allied countries—included lumber, fruit, whale oil, textiles, and foodstuffs. It wasn't military ammunition, but it was still vital to the Allied countries' supply chain, and its loss made an impact.

Germany only converted about nine ships into Sea Raiders, but they operated across the globe—in the Atlantic, Pacific, and Indian Oceans, and even the Tasman Sea near Australia and New Zealand. Their reach was vast, their tactics bold, and their crews highly trained.

However, the age of the Sea Raiders was short-lived.

By 1943, with the United States now in the war, Germany was under constant aerial bombardment. British Royal Air Force bombers struck at night, and the U.S. Army Air Forces bombed by day. These relentless air raids devastated Germany's industrial infrastructure, including the factories producing weapons, munitions, and ships.

Faced with the need to defend the homeland, Germany shifted its war priorities away from Sea Raiders. Resources were redirected toward building planes, tanks, and anti-aircraft defenses.

The last chapter came on October 17, 1943, when the final German Sea Raider was sunk by a U.S. Navy submarine.

Torpedoed beneath the waves, it marked the end of an era of stealth warfare at sea.

Sketch of a duel between a German U-boat just beneath the surface and an armed Allied merchant ship.

U-BOATS

Humans have long been fascinated by the ocean's mysteries—especially the hidden world beneath the waves. For centuries, inventors have dreamed of building machines to explore the underwater depths, and some of the earliest attempts go back farther than you might think—even as far as Alexander the Great, who was said to have descended into the sea inside a primitive diving bell.

But the first operational submarine—something closer to the underwater vessels we imagine today—was created in 1620 by Cornelis Drebbel, a Dutch inventor working for the English crown.

Drebbel's design was essentially a modified rowboat. To make it waterproof, he draped greased leather over the entire wooden frame, sealing it top and bottom. Oars poked out through the sides, fitted with watertight leather sleeves, allowing a 12-man crew to row the vessel underwater without letting water inside. It was basic, but it worked—Drebbel's submarine could submerge and navigate beneath the surface of the River Thames, amazing onlookers.

About 150 years later, during the American Revolutionary War, the United States developed its own version of an underwater vessel—called the Turtle.

The Turtle was a hand-powered, egg-shaped submersible designed to hold a single man. It was made mostly of wood and operated by foot pedals and hand cranks. What made it revolutionary was that it was the first submarine capable of independent underwater operation.

The Turtle could dive to a depth of about 25 feet—incredible for its time. By comparison, modern military submarines can descend to depths of 800 feet or more, using nuclear power and high-strength hulls.

Though primitive by today's standards, these early subs marked the beginning of a bold and dangerous new kind of warfare: beneath the waves.

Cutaway view of the Turtle.

During the American Civil War, both the Union and Confederate forces experimented with submarines.

The Union Navy developed the Alligator, while the Confederacy launched a more famous vessel: the CSS *H.L. Hunley*. The *Hunley* became the first submarine in history to successfully attack and sink an enemy warship—a Union ship called the USS *Housatonic*.

But the victory came at a terrible cost.

Shortly after the attack, the *Hunley* itself sank—taking its entire crew down with it.

Submarine warfare was proving to be effective … but dangerously unpredictable.

Submarines saw limited but important use during World War I—especially by Germany's growing U-boat fleet. However, it wasn't until World War II that submarines became a widespread and critical weapon of war, used by almost every major naval power.

You may have seen movies in which submarine crews in World War II lead exciting and adventurous lives—dodging depth charges and slipping silently through enemy waters.

But in reality, life aboard a submarine was incredibly difficult and often tedious—a word that means tiresome because of its length or repetitiveness.

Submarines patrolled underwater for weeks or even months at a time. And while the food was actually considered among the best in any branch of the military, it didn't stay that way for long.

Sub crews relied on foods with long shelf lives—like potatoes, ham, sausages, bread, and bacon. But without strong refrigeration (early submarines didn't have much of it), even these items spoiled quickly.

Food was stored everywhere—between torpedoes, under bunks, and even in the shower, which meant no one got to bathe. Imagine sweating all day in cramped, hot conditions … and never being able to shower.

Speaking of heat, temperatures inside these metal tubes often rose above 100 degrees Fahrenheit. Add to that the smell of sweaty men, stale food, diesel fuel, and engine grease, and you've got a pretty unpleasant atmosphere.

Privacy? Forget it. Unless you were the captain, you didn't get any personal space.

Because there weren't enough bunks for everyone, sailors slept in shifts, a system known as "hot bunking." One man would crawl into a bunk the moment another crawled out. Meanwhile, the engines roared, alarms blared, and crew members worked around the clock in tight quarters.

Even eating was done in shifts. And if the cook wasn't in the mood to make two hot meals, the second shift might be left with cold or lukewarm food.

It was exhausting, uncomfortable, and claustrophobic. But these harsh conditions were part of the price for operating one of the most dangerous—and strategic—weapons of the war.

Among the most notorious submarines of World War II were the German U-boats—short for *Unterseeboot*, meaning "undersea boat." These stealthy vessels terrorized Allied shipping in the Atlantic Ocean.

A torpedo being slid into a U-boat. Notice, too, the sub's deck gun mounted on the deck just below the torpedo, pointing out to sea.

The Germans hoped to use submarines to cut off the Allied countries from their supplies—especially food, fuel, weapons, and troops crossing the Atlantic Ocean.

When World War II began in 1939, Germany had about 57 operational submarines. But by the war's end, they had built a staggering 1,162 U-boats. These underwater predators were part of Adolf Hitler's plan to strangle Allied supply lines, especially those keeping Great Britain alive and fighting.

Britain, an island nation, relied on merchant convoys to bring in almost everything it needed. If those supplies were stopped, the country might starve—or be forced to surrender.

To carry out this mission, Germany created a special breed of sailor. U-boat crews were handpicked—the best and brightest, selected not just for their toughness but also for their technical and mechanical skill. They were part of the Kriegsmarine, the German navy, and underwent psychological testing to make sure they could handle the intense stress of underwater warfare: small, confined spaces, days of silence, and the constant fear of being hunted.

These crews were trained for what Hitler hoped would be the decisive weapon in the war—the U-boat.

In the early part of the war—from 1939 to 1942—German submariners experienced what they called "the Happy Times." That label didn't mean they were joyful undersea. Instead, it referred to the massive amount of destruction they inflicted on Allied shipping.

At first, Britain stood nearly alone against Nazi Germany. But on December 7, 1941, Japan attacked the United States at Pearl Harbor, Hawaii. The very next day, the U.S. declared war on Japan. And on December 11, 1941, Hitler declared war on America.

Suddenly, the conflict became a global war—and the tactics for submarine warfare evolved quickly.

Early in the war, German U-boats would surface and use their deck guns—large-caliber cannons mounted on top of the sub—to sink unarmed merchant ships. This made sense from the German point of view: why waste a valuable torpedo when they could just blast the ship apart from the surface?

There were other advantages, too. On the surface, a U-boat could travel at about 20 miles per hour—much faster and more maneuverable than when submerged. Underwater, U-boats were slow, crawling at a top speed of around nine miles per hour, and only for short periods before battery power ran low.

But as the Allies adapted—and new technology entered the war—the Germans were forced to rethink these tactics.

Surfaced German U-boat with crew. Check out the forward gun.

By 1941, German Admiral Karl Dönitz (pronounced "DER-nitz") introduced a new tactic that made U-boat warfare even more dangerous: the Wolf Pack strategy.

Here's how it worked.

When a single U-boat spotted a convoy—a group of merchant ships traveling together for protection—it would radio a signal back to German naval headquarters. That signal would summon multiple U-boats to the area. Together, they would coordinate a group attack, swarming the convoy like a pack of wolves—hence the name "Wolf Pack."

These German submarines often positioned themselves along known Allied shipping lanes, lying in wait. Once the slow-moving merchant ships came into range, the attack would begin—sometimes under cover of night, and often with devastating results.

Other tactics included laying underwater mines near harbor entrances, making it dangerous even for ships nearing safe waters.

To extend their time at sea, the Germans developed special U-boats called "Milk Cows." These weren't attack subs—they were giant, floating gas stations. Packed with diesel fuel, food, and torpedoes, they met up with U-boats in the middle of the ocean. Crew members connected long hoses to refuel the combat subs, allowing the attacking U-boats to stay active and avoid returning to port for supplies.

One of the U-boat's key defensive maneuvers was the "crash dive." If a submarine's lookout spotted an enemy plane or ship, an alarm would blare inside the sub. The sailors would rush forward to the bow (the front of the vessel), using their combined weight to help the submarine dive steeply and quickly, trying to escape before depth charges could be dropped.

As for torpedoes, German U-boats carried two basic types:

- Steam-powered torpedoes, which were faster but left a visible trail of bubbles, and
- Electric-powered torpedoes, which were slower but stealthier.

These weapons were about 23 feet long and 21 inches wide, with explosive warheads that carried more than 600 pounds of TNT. Once fired, they could travel over 20 miles per hour underwater—enough speed to punch a deadly hole in any ship they struck.

To maintain stealth—invisibility—U-boats usually surfaced at night to recharge batteries and ventilate the cabin. Before opening the hatch, the interior lights were switched to red. Why? Because red light preserved the crew's night vision, allowing them to see clearly when they emerged into the darkness.

But despite these clever and deadly tactics, the so-called "Happy Times" came to an end.

The Allies began to fight back—developing new technology, new tactics, and stronger teamwork that slowly turned the tide in the Atlantic.

German submarine sinking after being rammed by USS *Thomas* in the North Atlantic in 1944. Notice the mine chutes on the sub's foredeck.

During World War II, the Allies knew they had to find a better way to detect and destroy German submarines.

So the British created something called the Anti-Submarine Detection Investigation Committee—which had quite a long name! People shortened it by turning the acronym ASDIC into a nickname. In the United States, we simply called it sonar (short for "sound navigation and ranging").

Here's how sonar works:

An electronic device is lowered into the water. It sends out sound waves, kind of like an underwater "ping." If those sound waves strike a solid object, such as the metal hull of a submarine, the sound bounces back—just like an echo. That echo tells the technician listening with headphones the distance and direction of the object.

Believe it or not, sonar could sometimes even pick up the voices of crewmen inside the submarine if they were speaking loudly!

Once sonar—and radar, which detects objects in the air—became more advanced, the Allies could better locate U-boats sneaking through the Atlantic. That turned the tide in the battles of the Atlantic.

Back in World War I, the main way to attack submarines was with depth charges. These were heavy explosives, about 150 pounds each, made of cast iron. Sailors would roll them off the stern (the back of the ship) after estimating the depth of the submarine. Charges were usually preset to explode at depths such as 40, 80, or 100 feet.

But by World War II, German U-boats could dive as deep as 300 feet—well below the standard settings of those early charges.

So the Allies developed a better weapon: the Hedgehog.

Unlike the older method of dropping a single depth charge, the Hedgehog was a mortar launcher mounted on the bow (front) of the ship. It could fire 24 to 25 small bombs—each weighing around 65 pounds—in a wide circle around the submarine's estimated location.

Here's the best part: Hedgehog bombs didn't need to be preset for depth. They exploded only on contact with the submarine's hull.

Now imagine being inside a submarine while all that was happening. It would be like being trapped inside a steel drum while someone pounded it with sledgehammers. The echoing booms, shaking steel, and rising pressure would be almost unbearable.

And it got worse for the U-boats.

The Allies also began using aircraft carriers, which could launch planes to patrol over the oceans. These planes could spot submarines faster than ships, and often attacked them with bombs, rockets, or even depth charges.

To blend in with the sky, the undersides of the aircraft were painted light blue or white, making them harder to see from below. The longer it took a submarine to spot a plane overhead, the more time the pilot had to launch an attack.

Thanks to sonar, aircraft surveillance, and better weapons like the Hedgehog, the once-feared U-boats began to lose their edge—and the Allies started to win back control of the seas.

Aircraft attack a U-boat and a Milk Cow during World War II.

Around early 1943, the tide of the war at sea began to shift, thanks to a major breakthrough in codebreaking.

British intelligence had been working to crack the German naval code, which the U-boats used to communicate secretly. That code was created by a machine called the Enigma—a complex device that scrambled messages using a constantly changing pattern of letters and symbols.

At first, the code seemed unbreakable.

But then, the British got lucky. They captured a German U-boat and recovered an actual Enigma machine, along with logbooks that included daily codes and instructions. With this crucial information, British codebreakers, at a famous place called Bletchley Park, were able to crack the Enigma code.

Suddenly, the Allies could read secret German naval messages!

That meant they knew when a Wolf Pack was being formed and where it would strike. Armed with this knowledge, Allied ships could change their routes, avoid ambushes, or even prepare counterattacks.

Meanwhile, the Allies also started to arm their merchant ships more heavily. These supply ships, once easy targets, were now outfitted with deck guns, anti-aircraft weapons, and depth charges. These changes helped defend against not only U-boats but also German Sea Raiders.

The combination of radar, sonar, air patrols, better-armed ships, and the cracking of the Enigma code turned the tide in the Atlantic.

Germany had built over 1,100 U-boats during World War II. But they lost 753 of them.

Of the more than 40,000 German submariners who served, over 28,000 were killed, making it one of the most dangerous roles in the entire German military. That's a fatality rate of nearly 70 percent.

By the end of the war, what had once been one of Nazi Germany's most feared weapons had become a costly and crumbling force, outmatched by Allied technology, intelligence, and resilience.

BOOKS

Kriegsmarine: The Illustrated History of the German Navy in WWII by Robert Jackson (editor). Exploring everything from small coastal vessels to giant battleships, *Kriegsmarine* is a pictorial record of Germany's naval forces before and during World War II.

INTERNET

"How to make SUBMARINE with paper and cardboard": https://youtu.be/7g05QwGRpYU?si=2UXTTqWvHskT0NB3

"U-Boats (World War II)," a video from Simple History: https://youtu.be/94XkuKnpfDA?si=haUk8XI9Dt_PDKVf

"Hitler's Sea Wolves: The Granville Raid 1945," a short video with great historical footage of the last German naval raid of World War II—a sudden attack on the Normandy port of Granville, by a flotilla of German warships: https://youtu.be/s7szB5So5Gs?si=5nI7IS2wvKvW6i-l

MOVIES

History Rediscovered: Submarines at War (1946). This compilation features four rare World War II-era documentary films about submarines.

20,000 Leagues Under the Sea (1954)

Destination Tokyo (1943)

Mers El Kébir

July 3, 1940

French destroyer *Mogador* smoldering after shellfire at Mers El Kébir on July 3, 1940.

AFTER INVADING POLAND IN 1939—the event that launched World War II—Germany invaded France the following year.

In just six weeks, French resistance collapsed.

In June 1940, France signed an armistice with Germany. An armistice is a treaty that suspends fighting—but in this case, it was more like a surrender. The terms gave Germany control over northern France, including Paris, and handed over the French military, infrastructure, and resources to Nazi oversight.

The southern portion of France was technically still governed by the French—but it wasn't truly independent. That region, based in the spa town of Vichy, became a puppet government, cooperating with the Nazis. That territory became known as Vichy France.

While both zones suffered under Nazi occupation and repression, only one group of French patriots continued the fight for freedom. They were known as the Free French.

Led by a determined general named Charles de Gaulle, these brave men and women refused to accept Nazi control—or cooperate with Vichy leaders. Many of them fled to Britain, where they joined Allied forces and launched resistance efforts from abroad.

A World War II-era photo portrait of General Charles de Gaulle of the Free French Forces.

These freedom fighters were called the Resistance, and their leader was General de Gaulle.

After his country surrendered to Hitler, de Gaulle fled to England, where he established a Free French government-in-exile. From there, he encouraged French citizens to resist the

Nazis and the Nazi-collaborating Vichy regime. He also worked closely with the British to continue the fight for a liberated France.

But de Gaulle's forces were tiny, about 2,000 men stationed in England—hardly enough to challenge Hitler's powerful army.

World War II poster during German occupation of France, showing a citizen treading water and holding up his country's flag.

Meanwhile, Great Britain had a much bigger, more immediate problem with the German takeover of France: the French navy.

At the time, it was the fourth-largest navy in the world, equipped with battleships, heavy cruisers, destroyers, submarines, and thousands of skilled sailors. If Nazi Germany gained control of that fleet, Britain's survival would be in serious danger.

The Royal Navy was already spread thin defending against German and Italian fleets in the North Sea, the Atlantic Ocean, the Far East, and the Mediterranean. If Hitler added the French fleet to his war machine, he could potentially cut off all shipping routes into Britain and attempt an invasion of the British Isles.

Although Hitler promised not to use the French ships, Prime Minister Winston Churchill was deeply skeptical.

"But who in his senses would trust the word of Hitler after his shameful record and the facts of the hour?" Churchill said.

So Britain had to act. Fast.

France's main Mediterranean fleet was anchored at Mers El Kébir, a harbor in French Algeria, North Africa. Those ships—including modern battleships—were within striking distance of England's supply lines.

Churchill was direct: "At all costs, at all risks, in one way or another, we must make sure that the navy of France [does] not fall into the wrong hands and then perhaps bring us and others to ruin."

The French squadron at Mers El Kébir included:

- Two modern battleships: *Dunkerque* and *Strasbourg*, each about 26,500 tons (for context, the average modern car weighs less than one ton. These battleships were more than 20,000 times heavier!)
- Two older battleships: *Bretagne* and *Provence*, each about 22,000 tons
- A seaplane carrier: *Commandant Teste*
- Six destroyers.

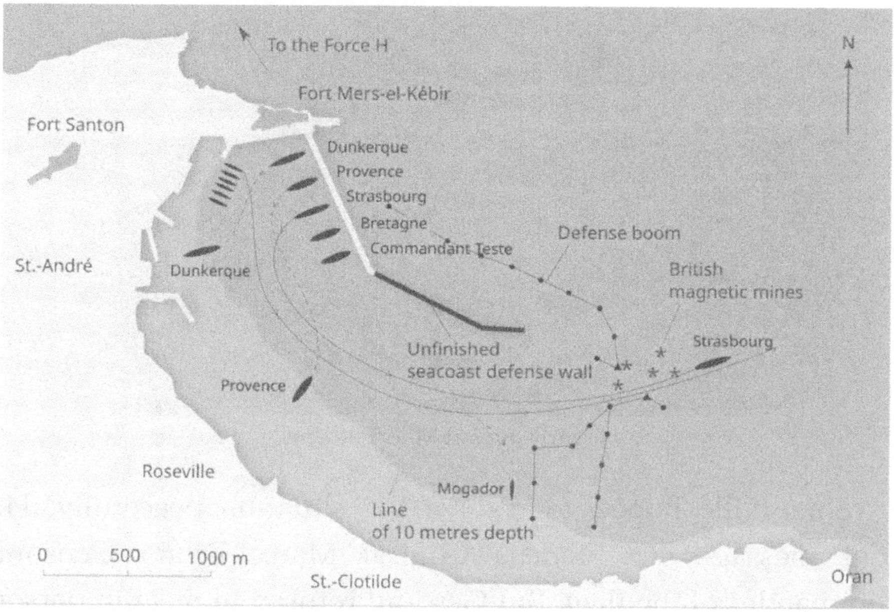

French ships docked at Fort Mers El Kébir. Also note the escape route to the right.

Churchill ordered Operation Catapult to neutralize the French fleet.

A British naval group, known as Force H, under the command of Admiral Sir James Somerville, was dispatched to carry out the plan.

But Somerville was a reluctant participant. After all, just weeks earlier, the French and the British had been close allies against Germany.

"You are charged with one of the most disagreeable and difficult tasks that a British Admiral has ever been faced with," Churchill told him. "But we have complete confidence in you and rely on you to carry it out relentlessly."

Somerville obeyed orders—duty before feelings.

On July 3, 1940, days after France's surrender, Force H arrived off the coast of Mers El Kébir.

French ships at Mers El Kébir.

Somerville hoped to resolve the situation peacefully. He sent messages to French Admiral Marcel-Bruno Gensoul, commander of the fleet. But Gensoul refused to meet in person and chose instead to communicate through messengers.

Churchill and the British presented four options to the French:

- Sail with the Royal Navy and continue the fight against Germany.
- Sail to a British port, where the ships would be impounded, and the crews sent back to France.
- Sail with reduced crews to Dakar, West Africa, where the ships would be immobilized.
- Scuttle—sink—the ships within the next six hours.

But if none these options were accepted, Force H was authorized to use "all means at your disposal" to prevent the French fleet from falling into German hands.

A French-speaking British officer, Captain Cedric S. Holland, was sent aboard the British destroyer HMS *Foxhound* to deliver the ultimatum. But Gensoul refused to meet Holland in person as well, saying he was too junior in rank.

As negotiations dragged on, British intelligence intercepted a French message suggesting the French might attempt to flee or even reinforce their position at Mers El Kébir.

There were also signs that the Germans were pressuring the French to resist.

"Settle this matter quickly," Churchill warned Somerville, "or you may have reinforcements to deal with."

Admiral of the Fleet Sir James Somerville.

Somerville gave his ships the order to prepare for battle.

At 5:15 p.m., French Admiral Gensoul revoked all peaceful options.

Somerville responded with one final warning: unless the French accepted terms, within minutes, the British would open fire.

At 5:55 p.m., Captain Holland returned to the *Foxhound*, and British ships laid magnetic mines across the harbor entrance to prevent escape (the location is noted on that map above).

Photograph of the bombardment in the harbor.

Then—BA-BOOM!

British battleships *Hood, Resolution,* and *Valiant* opened fire with their 15-inch guns, targeting the French warships.

The battle in the harbor.

Smoke billowed up, obscuring vision. Gunners had to use the Mers El Kébir lighthouse as a reference point for aiming fire.

A Swordfish biplane from a British aircraft carrier flew overhead, confirming hits on at least three French ships.

Then, a massive explosion shook the harbor as a British shell struck an ammunition magazine.

BA-BOOM!

The *Dunkerque* tried to escape, but struck a mine and was grounded near the shore.

The *Provence* was hit and beached.

The destroyer *Mogador* was blown nearly in half.

With limited room to maneuver in the small harbor, the French ships were sitting ducks. Within just 10 minutes, over 1,200 French sailors were dead or dying.

Somerville ordered a ceasefire, allowing any surviving crews to abandon their damaged vessels.

But the battle wasn't over.

At 6:20 p.m., a Swordfish pilot spotted the *Strasbourg* and five destroyers making a break for open sea.

The *Hood* gave chase but couldn't catch them.

Later that evening, Swordfish torpedo bombers found the *Strasbourg* and launched attacks from just 20 feet above the water. Despite damage, the ship escaped to a naval base in southeastern France.

Three days later, on July 6, British planes returned and sank the *Dunkerque*—along with an auxiliary ship.

And by then, almost 90 percent of France's operational battleship power had been neutralized.

The attack—though tragic—ensured the French fleet could never be used by the Nazis.

While the U.S. had not yet entered the war, President Franklin D. Roosevelt responded to the attack by sending 50 destroyers to support the Royal Navy.

The following year, America joined the war and fought alongside the British until Hitler's defeat in 1945.

Churchill would later reflect with sorrow on the decision to attack the French fleet at Mers El Kébir:

"This was a hateful decision, the most unnatural and painful in which I have ever been concerned ... The French had been only yesterday our dear allies ... But no act was ever more necessary for the life of Britain and for all that depended upon it."

WHO FOUGHT?

Winston Churchill.

A cannon blasted across the waves as a British Royal Navy ship sliced the sea like a steel shark. At the helm was a man with a fierce pale stare, a booming voice, and a brain full of brilliant ideas. He was Winston Churchill.

At age 36, Churchill commanded the most powerful navy in the world and became the First Lord of the Admiralty—the man in charge of the entire Royal Navy.

Churchill wasn't the type of leader who sat behind a desk. He charged into meetings like a battleship, wearing a suit and shouting things such as, "We need faster ships! Stronger guns!" Churchill helped design better ships, encouraged the use of airplanes flying from ships—later called aircraft carriers—and supported new military inventions for their time, such as tanks.

During World War I, Churchill came up with a bold plan—a secret mission to attack the enemy from behind, through a place called Gallipoli. This was a very risky and dangerous mission. And it failed (you can read about it in my *World War I* book).

Churchill accepted full blame for the failure.

Did he quit the navy?

Nope.

He put on a soldier's uniform and went straight to the front lines of the war—from giving orders to dodging bullets beside his troops.

Years later, when World War II broke out, the world needed heroic leaders to beat the Nazis.

Churchill returned to power—stronger and wiser for his failure. He used the lessons he learned to protect his country from danger and to help plan one of the most successful naval invasions in world history—D-Day.

"Never give in," he said. "Never, never, never!"

Winston Churchill lived by that saying.

Merchant Convoys of World War II

PQ-17

Arctic naval crews fighting the deadly deck ice.

YOU'VE NOW READ ABOUT many thrilling sea battles—ancient ships ramming into enemy hulls, barbed grappling hooks flying overhead to snag a vessel for boarding, cutlasses flashing in hand-to-hand combat, cannons roaring with deadly iron balls, and modern 15-inch turret guns thundering as enemy ships sank beneath the waves.

All of that warfare demands extreme courage.

But there's another kind of naval heroism—less celebrated but just as essential: the story of the merchant marines, especially those who served during World War II.

Technically, merchant mariners are civilians, not part of the military. Their primary job is to transport cargo across the globe. But in times of war, these brave sailors are called upon to risk their lives—sailing into combat zones, dodging enemy fire, and delivering critical supplies to troops who cannot fight without them.

During World War II, American merchant marines faced terrifying dangers: German warships, disguised Sea Raiders, lurking U-boats, and the brutal forces of nature—freezing storms, mountainous waves, and Arctic winds.

The North Atlantic Ocean's waters are some of the harshest on Earth. Waves can rise over 70 feet, and sea sprays freeze on deck, turning ships into floating icebergs. If a World War II merchant ship was struck and went down in these icy waters, survival chances were slim. Many seamen died from hypothermia within minutes.

At the start of World War II in 1939, Adolf Hitler—dictator of Nazi Germany—signed a non-aggression pact with Josef Stalin, dictator of the Soviet Union. The agreement, known as the Molotov-Ribbentrop Pact, stunned the world: the two opposing regimes promised to stay out of each other's way—and secretly agreed to carve up Eastern Europe between themselves.

But by summer 1941, Hitler had betrayed Stalin. In a massive surprise attack called Operation Barbarossa, Germany invaded the Soviet Union with more than three million troops, opening the brutal Eastern Front of the war.

Josef Stalin.

After that betrayal, Stalin and the Soviet Union joined the Allies in the fight against Nazi Germany.

But even though the Soviet Union was massive in size—because it included other Eastern European countries—it lacked the industrial capacity to produce enough tanks, planes, ammunition, food, and trucks to fight back effectively.

To keep the Soviet war effort alive, Britain organized naval convoys to bring supplies across the treacherous seas.

A convoy is a group of ships sailing together for mutual protection, often escorted by warships.

Even before the United States officially entered World War II—after the Japanese attack on Pearl Harbor in December 7, 1941—American merchant marines were already joining these dangerous convoys, delivering vital supplies to the Soviet Union under what was called the Lend-Lease Program. That program was an effort to help the Allies fight the Axis powers—which were Germany and Italy primarily.

These convoy missions—particularly the ones bound for northern Soviet ports such as Murmansk and Archangel—would become some of the deadliest naval supply routes of the entire war.

Convoy freighter blown up during an attack at sea.

The convoys supplying the Soviet Union had only two possible routes.

One option was to cross the Mediterranean Sea and sail into the Black Sea, eventually reaching southern Soviet ports. But this route stretched over 13,000 nautical miles and passed through dangerous waters controlled by Axis powers—including Italy and German-occupied territories.

It was long, exposed, and difficult to defend.

The other option was shorter—less than 2,500 nautical miles—but just as dangerous. This route led around the northern tip of Norway, through the Arctic Ocean, and into the Soviet ports of Murmansk and Archangel. The dangers were extreme: brutal weather, sub-zero temperatures, pack ice, and constant attacks by German U-boats, aircraft, and Sea Raiders.

Despite the hazards, the northern Arctic route became the primary supply line. Its shorter distance meant quicker deliveries, lower fuel costs, and reduced strain on already thin Allied resources. But the dangers were so great that Allied merchant ships were forbidden from stopping at sea, even for rescue operations. Any delay risked making the ships sitting ducks for German attacks.

Adolf Hitler was determined to choke off these convoys and starve the Soviet war effort. He ordered the full might of the Kriegsmarine (German navy) to the Arctic.

"Every ship that is not in Norway is in the wrong place," Hitler declared.

Imagine being a crewman on these convoys. Day and night, your ship rocks violently as 30- to 50-foot waves slam the hull. Water crashes over the deck—and then freezes, creating a thick coating of ice. If not chipped away quickly, that ice could make the ship top-heavy and cause it to capsize—flip over and sink.

But chipping ice on a pitching deck at sea is dangerous work. One wrong step, and you could fall overboard. In the frigid Arctic waters, survival time was often less than three minutes before the cold shut down your body completely.

In winter, the convoys sailed through endless darkness. Due to the polar night, the sun doesn't rise above the horizon for up to four months. The convoys hugged the coastline for navigation—but the coast was Norway, fully occupied by the enemy! German aircraft, submarines, and radar stations waited to strike.

You might think summer was easier. But in some ways, it was worse. In the "land of the midnight sun" the sun never fully sets, so ships were exposed to constant daylight, making them easier targets for Luftwaffe bombers, torpedo planes, and German surface raiders.

But the Allied merchant mariners stayed their course.

Convoy after convoy pushed northward, carrying thousands of tons of tanks, trucks, aircraft, ammunition, and fuel to support the Soviet fight against Nazi Germany.

The two ports—Murmansk and Archangel—became critical lifelines for the Eastern Front of the war. These supplies arriving through Arctic convoys helped the Red Army survive the darkest days of the war—and eventually push the German army back.

Map of the area where the merchant marines sailed to Russia. Notice Norway's coastline curving around to Russia and the ports of Murmansk and Archangel (on the map labeled "Archangel'sk").

Now let's dive into one particular convoy and its heroic battle to beat Hitler: PQ-17. (By the way, the convoys heading to the Soviet Union were labeled "PQ." The returning convoys were labelled "QP.")

PQ-17 sailed out of Iceland with 35 cargo ships on June 27, 1942.

Twenty-two ships were American, eight were British, two were Soviet, two were from Panama, and one was Dutch.

The convoy's protective escort was six destroyers, four corvettes, three minesweepers, four trawlers—large fishing boats—two anti-aircraft ships, and two submarines.

U.S. Navy Lieutenant Douglas Fairbanks Jr. was aboard one of the escort ships. Fairbanks described the convoy's departure from Iceland, heading for Archangel.

"Waddled out to sea like so many dirty ducks ... Everyone who was watching paid a silent tribute and offered some half-thought prayer."

Escorts and merchant ships of Convoy PQ-17 leaving Iceland for the Soviet Union.

But within days of departure, British intelligence learned that German forces had spotted Convoy PQ-17—and were preparing a massive assault.

German heavy cruisers, destroyers, and U-boats were reportedly on their way to intercept. But also coming was the *Tirpitz*, the pride of the German navy (which you will read about in the next chapter). The mere presence of the *Tirpitz* was enough to terrify any Allied sailor.

Realizing the entire convoy was now at risk, British naval command issued a series of urgent, top-secret orders:

"Most immediate and secret," read the first signal.

"Cruiser force withdraw to the westward at high speed."

This meant that the convoy's heavy escorts—its strongest warships—were pulling out!

A second, even more devastating order followed:

"Immediate. Owing to threat of surface ships, convoy is to disperse and proceed to Russian ports."

And then came the most chilling message of all: "Most immediate. Convoy is to scatter."

Scatter.

With that single word, PQ-17 was abandoned to its fate.

The idea for scattering was to avoid presenting the Germans with a large, easy target. British leaders feared that if the *Tirpitz* attacked, it could sink all the merchant ships in one horrific battle. By scattering, they hoped a few vessels might escape and survive.

The convoy's defenseless merchant ships were to sail alone through some of the most dangerous waters in the world—hunted by submarines from below, bombers from above, and surface raiders on the horizon.

The sailors aboard PQ-17 suddenly found themselves utterly alone, with little hope of help—and hundreds of miles left to go.

The U.S. Navy cruiser USS *Wichita* (CA-45) and HMS *London*, part of the military escort for PQ-17.

A final message crackled over the radio waves:

"Escort to merchant ships. We are sorry to leave you like this … Good luck … Looks like bloody business."

It was a stark, heartbreaking admission. For the sailors aboard the merchant vessels of Convoy PQ-17, the British order to "scatter" was nothing short of a death sentence.

Now alone in hostile Arctic waters, the merchant ships could be picked off one by one by German U-boats, aircraft, and surface raiders. They had already completed more than half the journey—only a few hundred miles remained to reach the Soviet ports of Murmansk or Archangel. But now, instead of traveling as a protected fleet, the ships were spread out and exposed across a vast and dangerous ocean.

Soon, the merchants were radioing horrific distress calls:

"Am being bombed by a large number of planes."

"On fire in the ice."

"Abandoning ship."

"Six U-boats approaching on the surface ..."

The merchant vessel *Washington* carried 600 tons of high explosives—a floating bomb—when German Stuka dive bombers swooped down and riddled the ship with gunfire and bombs. The attack started a massive fire that threatened to detonate in a catastrophic blast.

The crew had no choice but to abandon ship. They scrambled into lifeboats and rowed into the icy seas, watching helplessly as their ship burned in the Arctic gloom.

Soon, another ship appeared—a British vessel trying to mount a rescue. But the men in the lifeboats waved it off, shouting for it to turn away. They knew what would happen.

And they were right.

The rescue ship was torpedoed and sunk, another casualty in the unfolding disaster.

The lifeboat crew of the *Washington* refused to give up. With ingenuity and grit, they rigged sails from scrap canvas and navigated the freezing waters for 10 harrowing days.

Incredibly, they reached safety in the Soviet Union, frostbitten and exhausted—but alive.

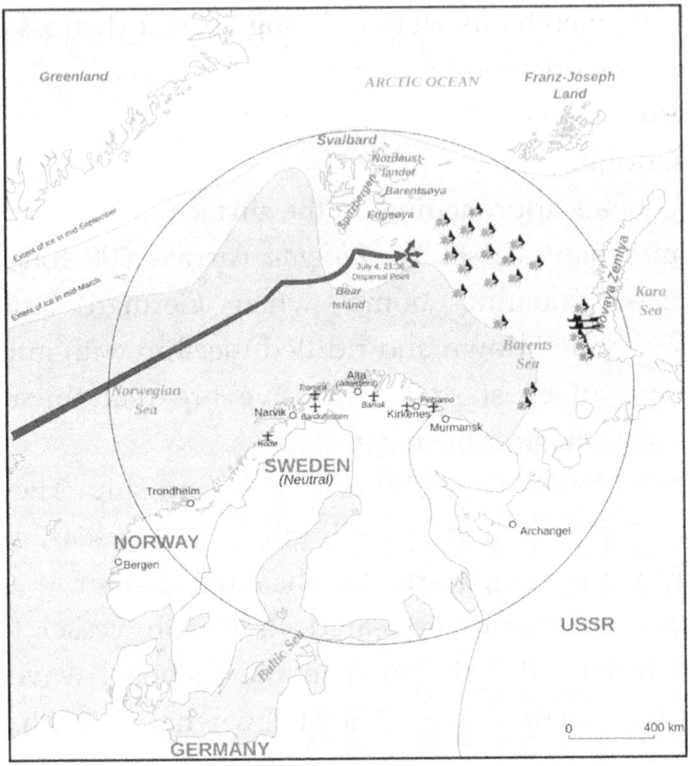

Approximate positions of the sunken PQ-17 ships.

To escape German attacks, the British escort trawler *Ayrshire* took bold action. Its captain led three merchant ships northward—into the Arctic ice.

With enemy U-boats and bombers hunting the scattered convoy, the *Ayrshire*'s captain gambled on nature itself for protection. The ships steamed into frozen waters, cutting their engines once surrounded by thick ice floes.

There, surrounded by silent glaciers and drifting pack ice, the crews began to camouflage their ships. They painted the hulls white to blend in with the frozen landscape and covered the decks in white linen. They even repositioned the Sherman tanks strapped to the decks, arranging them in defensive formation—their main guns loaded and pointed outward in case of an ambush.

Then they waited—motionless in the frozen silence, hoping to escape detection.

Eventually, when no German reconnaissance aircraft spotted them, the ships restarted their engines, slipped out of the ice, and continued toward their destination.

All four ships made it safely to the Soviet Union.

But elsewhere, the scattered remains of Convoy PQ-17 were under relentless attack.

On July 5, German aircraft struck again, sinking six more Allied ships. That same day, German U-boats sank another six.

Over the next five days, the attacks continued—nine more merchant ships were destroyed.

PQ-17 was becoming a catastrophe. Once a well-organized convoy of 35 merchant ships, it was now being annihilated, piece by piece, in one of the worst naval disasters of World War II.

German U-boat flying victory pennants and the flag of a captured PQ-17 merchant ship.

Of the 35 merchant ships that had left Iceland as part of Convoy PQ-17, 23 were sunk.

Only nine ships reached the Soviet port of Archangel—two British, four American, one Panamanian, and two Russian.

These survivors delivered less than half of their intended supplies. The losses were staggering:

- 3,350 vehicles
- 210 aircraft
- 430 tanks
- And approximately 100,000 tons of other cargo—ammunition, food, fuel, and other military necessities.

These devastating losses, combined with the high number of casualties, forced the Allies to suspend Arctic convoys for a time.

But by the autumn of 1942, with the United States now fully engaged in the war, the Allies resumed their mission. A new convoy, PQ-18, set sail for the Soviet Union. This time, it was protected by a formidable escort: 53 Allied warships, including submarines and even an aircraft carrier.

And still, the Germans struck. Thirteen more ships were sunk.

Despite these dangers, the Arctic convoys continued—relentlessly—until the end of the war in 1945.

In total, 1,526 merchant ships successfully arrived in the Soviet Union, carrying vital war supplies that changed the course of the war for the Soviet people.

In 1941 alone, amid deadly attacks from German aircraft, U-boats, and Sea Raiders, the convoys delivered:

- 750 tanks

- 800 airplanes
- And more than 100,000 tons of military cargo.

These deliveries played a critical role in turning the tide of war on the Eastern Front.

Yet the cost was steep.

In PQ-17 alone, 153 merchant mariners were killed.

During the war, about 100 Allied supply ships were destroyed in the Arctic convoys.

Unbelievably, the Soviet Union never formally thanked these brave civilian mariners.

The Soviets argued that these crews weren't "real" soldiers—they weren't firing weapons on the battlefield or flying fighter planes in combat.

But history makes one thing clear: without the merchant mariners, many of them American, the Soviet Union would have lacked the tools and materials to withstand the German onslaught.

In other words, a person doesn't have to hold a gun to help win a war.

In Britain, the full truth about PQ-17 was kept from the public at the time. Government officials knew they had ordered the convoy to scatter, leaving the civilian crews defenseless. There was no excuse for the abandonment. Military planners had failed in their duty.

Later, Prime Minister Winston Churchill admitted, "All risks should have been taken in defense of the merchant ships."

Churchill also called PQ-17 "one of the most melancholy naval episodes in the whole of the war."

In total, Allied merchant fleets delivered an estimated 45 million tons of supplies during World War II.

A very large share of that tonnage went to the Soviet Union—without formal recognition or gratitude.

WHO FOUGHT?

Douglas Fairbanks Jr. was a reserve officer in the United States Navy and served on the cruiser USS *Wichita* during the disastrous PQ-17 operation.

While serving abroad in England, Fairbanks saw the training of British Commandos in raiding parties, diversions, and deception operations. He also trained with the unit and participated in several "harassment raids." That work gave him a deep appreciation for the military art of deception.

Fairbanks soon convinced the U.S. Navy of its need for a similar elite unit of men who were trained to conduct tactical diversion and deception missions.

The Navy recruited 180 officers and 300 enlisted men for what became known as the Beach Jumpers. Candidates for the program were required to have no seasickness, a lot of experience handling small boats, enough electrical knowledge to repair a home radio, and a basic understanding of celestial navigation—piloting a ship based on the positions of the stars.

"The Navy is requesting volunteers for prolonged, hazardous, distant duty for a secret project," the recruitment leaflet read.

The "Beach Jumpers" name came from their ability to quickly hit a beach and confuse the enemy with harassment and deception—basically scaring the enemy into confusion.

The identities and activities of the Beach Jumpers were very highly classified. Any leaked information could ruin even the most cunning deceptions.

Beach Jumpers used specialized deception equipment to make the enemy believe a small group of Beach Jumpers was a 70,000-man amphibious landing force—when in actuality, that force was landing a great distance away.

Beach Jumpers trained in the handling of small boats, seamanship, gunnery, demolition, pyrotechnics (blowing things up), and meteorology. They used 63-foot double-hulled plywood boats manned by an officer and a six-man crew. The boats came equipped with twin .50-caliber machine guns, small rockets, smoke pots, generators, and floating time-delay explosive packs.

Beach Jumpers also carried specialized deception equipment, such as recorders, amplifiers, and naval balloons capable of radar detection.

The Beach Jumpers' first action was in Operation Husky, with the Allied invasion of the Italian island of Sicily. The Beach Jumpers continued to conduct hazardous shallow-water operations throughout the Mediterranean for the rest of World War II.

Lieutenant Commander Fairbanks was awarded the United States Navy's Legion of Merit with bronze V (for valor) and the Silver Star, also for valor, for planning these diversion-deception operations and for his combat service.

After World War II, Fairbanks returned to his prior work starring in Hollywood movies. (He was the only son of another famous actor, Douglas Fairbanks Sr.)

Among Fairbanks Jr.'s movies are *Sinbad the Sailor, The Exile, The Prisoner of Zenda, Gunga Din,* and *The Corsican Brothers.*

BOOKS

Convoy PQ-17 1942: Disaster in the Arctic by Angus Konstam. A gripping account of the most famous convoy operation of the war, which marked a high-water mark for the German naval campaign in the Arctic.

INTERNET

"PQ-17 – A Convoy Scatters ... To Their Doom." Excellent historic footage. https://youtu.be/HGe78WMTTmM?feature=shared

MOVIES

Sinbad the Sailor (1947)

Gunga Din (1939)

The German Battleship Tirpitz

1939–1944

AMONG THE MANY DANGERS faced by the Arctic convoys, there was one ship—one towering, steel-clad menace—that haunted Allied naval planners like a ghost in the fog.

Her name was the *Tirpitz*.

Sister ship to the famed *Bismarck* (described in my *WW2 in Europe* book), the *Tirpitz* was the largest battleship ever built by Germany—and the heaviest battleship ever constructed by a European navy.

Her mere presence in the fjords of Norway forced the Allies to reroute convoys, commit massive resources to surveillance and containment, and live with the constant fear that this floating fortress might leave her harbor and wreak havoc on the fragile supply lines to the Soviet Union.

At one point, just the possibility that the *Tirpitz* might attack Convoy PQ-17 contributed to the catastrophic decision to scatter the convoy—leading to the loss of nearly two-thirds of the ships.

That's the power the *Tirpitz* held, not just as a battleship but as a strategic weapon.

The *Tirpitz* being launched. Look at the people on its deck and on the dock. They look like ants compared to this massive battleship.

Hitler called his new navy the Kriegsmarine (pronounced "kreegs-marine"), a German word that translates to "war navy." He envisioned it growing as powerful as Britain's Royal Navy, which at that time was the most formidable naval force in the world.

Hitler's ambitious naval plan included:

- 10 full-sized battleships
- 15 "pocket battleships" (smaller, faster versions with heavy guns)
- 4 aircraft carriers
- 250 submarines
- And more than 100 cruisers and destroyers.

But there was a problem.

The Treaty of Versailles, which ended World War I, specifically forbade Germany from building submarines and limited the size of its navy.

Hitler ignored those restrictions.

Secretly, throughout the 1930s, he ordered the rapid construction of warships, submarines, and naval bases—all while pretending to respect the treaty's terms.

As history would prove again and again, Hitler could never be trusted.

The *Tirpitz* at sea. Check out all its guns.

And yet, for all its immense firepower, here's the most astonishing part: the *Tirpitz* never really saw direct combat.

Why?

Because she was so expensive to build that the Germans were terrified of losing her in battle!

On the other side, the Allies were desperate to destroy her.

This giant battleship was such a threat—even when she wasn't fighting—that the Allies would end up launching more than 20 military operations just to take her out.

The *Tirpitz*'s first mission came in the Baltic Sea. The goal?

Prevent the Soviet navy from leaving port in Leningrad.

It worked—the Soviets were so intimidated by the *Tirpitz* that they kept their ships anchored!

This massive battleship didn't fire a shot—her mere presence was enough to stop a fleet.

Her next deployment, in January 1942, followed a similar strategy. This time, she was sent to the coast of Norway. But this wasn't an attack mission—it was a diversion.

The Germans wanted the British to panic and send warships north to chase the *Tirpitz*. Because while the British were distracted, the Kriegsmarine could slip into the Atlantic Ocean and attack the Allied shipping lanes—cutting off vital supplies to England.

In truth, the *Tirpitz* wasn't built to go bow-to-bow in ship-to-ship combat. Her purpose was psychological warfare. She was meant to scare the Allies into reacting, rerouting, and spending resources they couldn't spare.

But that plan came at a cost for the Germans.

The Allies were so obsessed with destroying the *Tirpitz* that the Germans had to work constantly to hide her.

But ... how do you hide a battleship that's more than two football fields long?

The answer: the Norwegian fjords (pronounced "fee-yords").

Fjords are narrow waterways surrounded by towering cliffs and steep mountain walls.

The Germans would dock the *Tirpitz* right up against one of these rocky slopes—and then chop down nearby trees and shrubbery to camouflage her from Allied pilots flying overhead.

From the sky, she could look like part of the mountain.

The *Tirpitz*, camouflaged by trees and shrubbery.

Another method of concealment was fake fog.

Seriously!

The Germans mixed water and chloro-sulfuric acid, then blew the thick mist toward the ship, cloaking it from aerial view.

Fake fog concealing the *Tirpitz* in Norway.

However, on September 8, 1943, the *Tirpitz* finally went into action.

Her gigantic guns thundered across the sea, bombarding Allied forces as they landed on a remote island north of Norway. It was the first time this massive battleship had fired in actual combat.

That was all the British needed to see.

They immediately began organizing concentrated missions to destroy the *Tirpitz* once and for all.

Their first move was brilliant: destroy the only harbor facility capable of repairing the *Tirpitz*. Without specialized equipment, even minor damage to the ship would become catastrophic. This meant every future hit could potentially be permanent.

The British launched a series of air attacks directly on the battleship—daring bombing raids and new technologies developed just for this purpose.

Among the most innovative weapons were Chariot torpedoes.

Imagine this: engineers took actual torpedoes, mounted seats on them, and trained soldiers to ride the torpedoes underwater, steering them as close as possible to the *Tirpitz*.

Once there, the men would detach explosive charges—and try to escape before the blasts went off.

It was underwater sabotage of the most dangerous kind.

Chariot torpedo without its warhead. The rider wears an oxygen mask and underwater suit.

However, none of these early attacks succeeded. The *Tirpitz* remained afloat—and dangerous.

So the British devised another plan.

This time, they built miniature submarines with a daring objective: sneak into the *Tirpitz*'s harbor and attach explosive mines directly to the battleship's hull.

Each mine had a delayed fuse, giving the mini-sub crews just enough time to escape before the detonation.

But on the day of the planned attack, disaster struck.

A brutal storm rolled in, churning the sea into chaos. The mini-subs couldn't risk entering the harbor. The mission had to be scrubbed.

Instead of giving up, the British decided to take the idea one step further.

They introduced a new generation of mini-submarines called X-Craft.

The X-Craft were about 50 feet long, crewed by three sailors and one diver, and specially designed for high-risk sabotage missions. They were towed by full-size "mother submarines" into operational waters.

At the launch point, the X-Craft crews transferred aboard, detached from the tow, and piloted the small vessel toward their target.

After the attack, the plan was to rendezvous with the mother sub and hitch a ride back to safety.

But the X-Craft faced trouble, too.

On its way to the *Tirpitz*, one of them sank, taking its entire crew to a watery grave.

Then, on September 20, 1943, opportunity finally arrived …

X-Craft underway.

Several X-Craft mini-subs were stationed off the Brattholm Islands, located along the rugged coast of northern Norway, where the *Tirpitz* was anchored in a nearby fjord.

The German harbor was heavily protected. Steel anti-submarine nets stretched from the water's surface all the way to the sea floor, forming a deadly underwater cage. These nets were only opened when authorized German ships or subs were expected.

But then—an opening!

One X-Craft, commanded by Lieutenant Donald Cameron, spotted a German ship approaching the harbor. Cameron made a bold decision: he maneuvered his X-Craft directly behind the enemy vessel, hiding in its wake as it passed through the opening in the net.

It was an incredibly risky move—but it worked. The British mini-sub slipped into the harbor, undetected.

But almost immediately, disaster struck.

The X-Craft's compass failed—suddenly, the crew had no sense of direction. Then, even worse, the periscope broke, blinding them entirely to their surroundings.

Now deep inside enemy waters, they were effectively navigating blind—unable to see or steer accurately.

And yet, the crew pressed on …

Mini-sub crewman during World War II. Not much room inside, is there?

But imagine the fear.

Sailing blind, underwater, deep inside an enemy-controlled port—completely unsure of your position—knowing that at any moment, the enemy could open fire and blow you out of the water.

And then, things got worse.

Lieutenant Donald Cameron's mini-submarine suddenly struck a submerged rock. The jolt was terrifying. The crew had no idea how serious the damage was—but if the hull had cracked, the pressure at that depth could crush them, or a leak could fill the vessel with freezing seawater.

They needed to check for damage. And there was only one way to do that: they had to surface.

But surfacing inside a German harbor? Surrounded by enemy guns? It was as close to suicide as it gets.

Meanwhile, aboard the *Tirpitz*, a junior German officer spotted something unusual in the water. He rushed to report it—he had seen what looked like the outline of a submarine.

But his superior officer just laughed.

"It was probably just a porpoise," he said dismissively.

Below the surface, Cameron and his crew—struggling to survive—had no choice. They surfaced the damaged X-Craft, knowing it might be the end.

To prevent the Germans from capturing this secret mini-sub, Cameron immediately gave the order to scuttle it—he opened the internal water valves and flooded the vessel.

As it sank, German forces captured the crew. They would spend the rest of World War II as prisoners of war.

But while Cameron's sub was caught, another X-Craft had slipped through the German defenses undetected. That crew made it all the way to the *Tirpitz*.

With expert precision, they maneuvered along the port (left) side of the battleship. One of the crew—a diver—exited the X-Craft and attached massive explosive charges underneath the keel (the keel is the long, central structural base of a ship that runs along its bottom, like a spine).

But just as the mini-sub tried to escape, it became entangled in the netting that surrounded the harbor.

The delay proved deadly.

German sailors aboard the *Tirpitz* spotted the sub. Machine guns opened fire. Hand grenades splashed into the water. The mini-sub was under full attack.

Why didn't they just use the *Tirpitz*'s massive 15-inch guns?

Because those gigantic weapons were designed for long-range sea battles. They couldn't aim low enough to fire at something so close and so small.

As the X-Craft began to sink from the damage, the commander raised a white flag in surrender. But it was too late for two members of the crew—they went down with the sub.

The other two were captured and taken prisoner.

One final X-Craft was still outside the harbor. But as it approached, it was spotted—and this time, the big guns of the *Tirpitz* opened fire.

BOOM!

The mini-sub was hit and completely destroyed—no survivors.

But then—BOOM!

Two massive underwater explosions rocked the harbor. The explosive charges planted under the *Tirpitz* detonated with such force that they lifted the 50,000-ton battleship several feet out of the water!

The damage was significant.

The ship's propellers were mangled. The rudders were warped. Internal systems suffered serious disruption.

But—astonishingly—the *Tirpitz* didn't sink.

Over the next six months, German engineers worked tirelessly to repair her. Eventually, the ship sailed again.

But by that time, the war had begun to shift.

The Nazis were no longer on the offensive—and the very nature of naval warfare was changing.

Aircraft carriers now ruled the seas, launching fighter planes from miles away—far outside the range of even the *Tirpitz*'s enormous guns.

Fighter pilots aboard an Allied aircraft carrier study maps of the *Tirpitz*'s location.

In April 1944, the era of airpower finally caught up with the *Tirpitz*.

The British Royal Navy launched 40 fighter-bombers from an aircraft carrier, flying them straight toward the battleship's hidden position in the Norwegian fjords.

The results were devastating: 14 bombs struck the *Tirpitz*—tearing into her armored deck, gutting internal compartments, and starting fierce fires. More than 120 crewmen were killed. Another 316 were wounded.

But even after all that destruction … the beast did not sink.

Like some kind of cursed sea monster, the *Tirpitz* refused to die.

In August 1944, the British returned with another airstrike. More bombs fell. More damage was done.

And yet again—the *Tirpitz* stayed afloat!

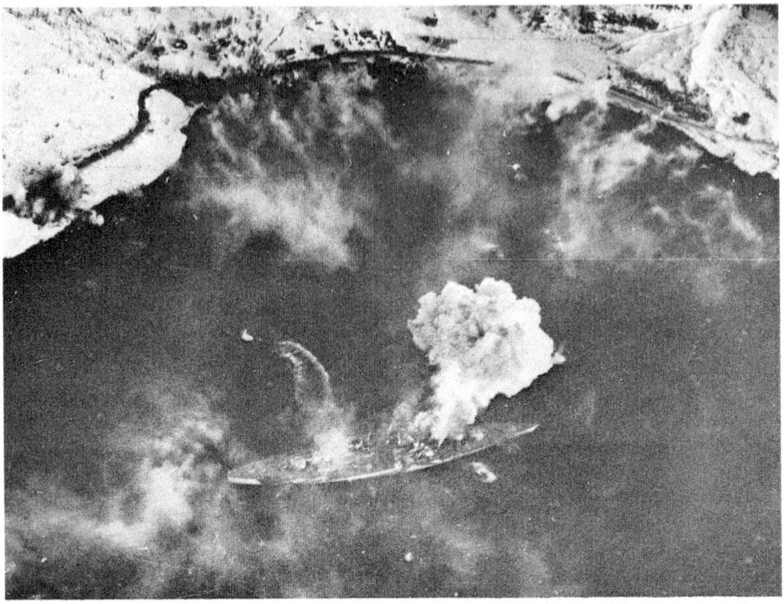

The *Tirpitz* under air attack.

The following month, the British launched yet another assault—this time with a new level of force.

They sent 27 Avro Lancasters, massive four-engine heavy bombers usually tasked with leveling entire cities. But their mission wasn't a city—it was a single ship.

This time, the bombers carried Tallboy bombs—also known as "earthquake bombs." Each bomb weighed a staggering 4 tons and was designed to pierce deep into the ground—or, in this case, deep into a warship—and detonate with devastating force.

The Lancasters dropped their Tallboys directly on the *Tirpitz*.

And still … the ship did not sink.

But this time, it was wounded.

Badly.

The *Tirpitz* limped to a repair dock. It was no longer seaworthy—its days as a mobile battleship were over. The German navy now planned to use it as a floating artillery platform, stationary but still dangerous.

The British, however, were not finished.

They returned—this time with 32 Avro Lancasters, fully loaded for another attempt to finish what they had started.

The bombers struck again. More destruction. More fire.

And still … the *Tirpitz* stayed afloat.

British planes dropping bombs on the *Tirpitz* below.

Then, in November, the British returned one last time.

They sent another 32 Avro Lancasters, each one armed again with the devastating Tallboy bombs.

This time, the bombs found their mark.

Two direct hits, followed by four near-hits, struck the *Tirpitz* with crushing force. Steel buckled. Bulkheads shattered. The behemoth began to bleed water.

Within minutes, the mighty battleship started to list—tilt to one side—then capsize.

Fifty thousand tons of iron, firepower, and twisted ambition rolled over and plunged beneath the waves.

The *Tirpitz* was gone.

The sinking claimed the lives of 971 German sailors—a tragic and violent end to one of the most feared warships ever built.

The *Tirpitz* capsized.

The wreck of the *Tirpitz* remained in the icy waters of Norway until after World War II. In 1948, a German-Norwegian salvage company began the slow process of dismantling the once-mighty battleship, selling off her steel and fragments for scrap.

It was an inglorious end for the crown jewel of the Kriegsmarine.

One historian later reflected:

"The famous battleship's most valuable contribution to the war effort was her mere existence. ... Her potential posed such a threat that Britain injected terrific resources into effecting her destruction."

The *Tirpitz*—which had once loomed as a symbol of German naval power—was ultimately undone not by a great sea battle, but by the fear she inspired. That fear alone diverted thousands of Allied troops, dozens of warships, and scores of bombers across years of missions—just to eliminate the threat she might pose.

In the end, she was a ghost ship of war, sunk by the weight of her own legend.

WHO FOUGHT?

Admiral Alfred von Tirpitz.

Born in 1849, Alfred von Tirpitz grew up fascinated by the ocean. At age 16, he joined the German navy and rose through the ranks and worked hard, rising to become one of the most influential admirals in German history.

During the late 1800s, Tirpitz convinced then-German emperor Kaiser Wilhelm II to build a fleet of massive warships called dreadnoughts. These ships were gigantic metal monsters bristling with cannons. (And a prelude to the massive ship built in his name.)

With Tirpitz in charge, Germany's navy grew faster than ever before. That hyper-growth was part of a Tirpitz plan called Risk Theory. If Germany's navy grew big enough, Tirpitz believed that Britain would be too scared to fight it.

But Germany's massive shipbuilding soon launched a naval arms race, with both Germany and England racing to build the biggest, baddest fleets.

When World War I erupted in 1914, all that sea power was put to the test. Germany and Britain faced off in ocean battles, although most of the giant ships stayed hidden in port.

Tirpitz wanted to fight even more, but others disagreed. In 1916, he stepped down from the navy.

Though famous for large warships, Tirpitz also helped invent Germany's U-boat strategy, leading to the subs becoming among the most feared weapons in both World Wars.

After World War I, Tirpitz became a politician and wrote his memoirs, in which he criticized many of his country's leaders.

His dream of making the German navy into the world's greatest sea force was never realized.

BOOKS

Sink the Tirpitz 1942–44: The RAF and Fleet Air Arm duel with Germany's mighty battleship by Angus Konstam (author) and Jim Laurier (illustrator). With detailed maps and diagrams, this short book explains the aircraft and ordnance that the British used to attack the *Tirpitz*, the evolving strategies, and why the task was so difficult.

INTERNET

Newsreel about the life and death of the *Tirpitz* and aerial attacks on her: https://web.archive.org/web/20110611072431/http://www.britishpathe.com/record.php?id=23512

The Battle of the Philippine Sea

June 19–20, 1944

Grumman F6F-3 fighter landing aboard the carrier USS *Lexington*, the flagship of Task Force 58, during the Battle of the Philippine Sea, June 1944.

IN THE MID-1940S, an argument erupted over which naval vessel was superior: aircraft carriers or battleships.

Aircraft carriers had the power to launch devastating attacks on enemy ships without ever needing to see them directly.

Battleships, on the other hand, boasted massive firepower capable of blasting enemy vessels out of the water.

That debate was settled by the largest carrier-versus-carrier battle in history: the Battle of the Philippine Sea.

It involved 24 aircraft carriers and more than 1,000 carrier-based aircraft.

During World War II in the Pacific, the Allied naval strategy became known as "island hopping." The Pacific Ocean spans more than 60 million square miles and contains thousands of islands. Rather than invading every Japanese-held island, Allied forces focused on key targets—particularly those with Japanese airbases or strategically important locations. (You can read more about this strategy in my *WW2 in the Pacific* book.)

By 1944, the U.S. Navy was closing in on the Central Pacific, specifically the Japanese-held islands of Saipan, Guam, and Tinian. Capturing these islands would cut off Japanese supply lines and give the Americans bases from which they could launch bombing raids on the Japanese homeland.

U.S. Navy "Seabees" stacking bombs for future use by the Army and Navy planes based in the Philippines, March 1944. Note the small, motorized crane for bomb handling.

Realizing this island-hopping threat, Japan responded with a desperate counteroffensive.

Admirals Soemu Toyoda and Jisaburō Ozawa were sent to intercept the advancing U.S. fleet in an operation codenamed A-Go.

The Japanese fleet included:

- 5 aircraft carriers
- 4 light carriers (carrying fewer aircraft)
- 5 battleships
- 11 heavy cruisers
- 2 light cruisers
- 29 destroyers
- And approximately 500 aircraft.

They faced the U.S. Fifth Fleet, specifically Task Force 58, led by Admirals Raymond Spruance and Marc Mitscher. The American force was enormous, including:

- 15 aircraft carriers
- 7 battleships
- 21 cruisers
- 8 light cruisers
- 69 destroyers
- And nearly 1,000 aircraft.

The Japanese believed they had a key advantage: land-based aircraft stationed on nearby islands. These planes could fly longer distances than American carrier-based aircraft. But that edge vanished when American planes began systematically bombing Japanese island airfields.

American submarines patrolling near the Japanese home islands had already warned the fleet: the Japanese navy was on the move.

Japanese Admiral Jisaburō Ozawa.

On June 19, 1944, Admiral Ozawa launched his first air-strike—69 fighter aircraft—to strike the U.S. fleet.

That attack marked the opening of the Battle of the Philippine Sea.

The U.S. responded with 45 F6F Hellcats, blue-colored fighters known for their ruggedness, speed, and firepower.

The Hellcat's specifications are awe-inspiring:

- Top speed: 386 mph
- Ceiling: 37,000 feet
- Wingspan: 42 feet, 10 inches (not as wide as a football field, which is 160 feet across, but still very large for a carrier-based fighter)
- Bomb payload: up to 2,000 pounds

- Armament: six .50-caliber machine guns.

Pilots nicknamed the Hellcat "The Big Blue Blanket" because of its protective presence in the sky. It was also about 50 mph faster than Japan's A6M Zero, though the Zero had better maneuverability and carried two 20mm cannons.

The Hellcat's advantage was durability. It could take hits and keep flying, while the Zero had no armor and no self-sealing fuel tank—making it extremely vulnerable when shot.

Hellcat night fighter circa 1944.

The American pilots knew how to defeat the Zero.

In 1942, a Zero crash-landed in the Aleutian Islands of Alaska. The relatively intact plane was shipped to the mainland (Alaska wasn't yet a state). Military engineers studied it carefully and discovered its weaknesses: no pilot armor, no self-sealing fuel tank, and weak construction.

U.S. pilots were taught to dive from above and behind to target the Zero's unprotected cockpit and fuel tank.

Back to the battle.

That first Japanese strike was repelled by the Hellcats, who shot down many of the 69 planes. Then Ozawa launched a second wave—127 aircraft.

American defenses again sprang into action. Of those 127 planes, 99 were shot down—many exploding in fireballs.

Ozawa still didn't give up.

He launched two more massive waves of aircraft. But they met the same fate: slaughter in the skies.

Japanese Zeros flying during World War II.

By the end of June 19, Japan had lost 294 planes, including many launched from both carriers and land bases.

The U.S. had lost only 22 fighters.

The Japanese were now suffering from another critical disadvantage: lack of experienced pilots. Many of their best aviators—including those who bombed Pearl Harbor—had been killed at Midway and in other battles since 1942. Their replacements were poorly trained and inexperienced in combat.

Meanwhile, below the waves, the U.S. submarine USS *Albacore* fired torpedoes at Ozawa's flagship, the *Taihō*, causing fires that led to a massive explosion. (The *Taihō* sank, but Admiral Ozawa survived—he was transferred to another ship.)

Another submarine, the USS *Cavalla*, struck and sank the carrier *Shōkaku*—a Japanese carrier that had also taken part in the Pearl Harbor attack.

Japanese Carrier Division 3 under attack by Task Force 38 planes, June 20, 1944. The battleship in the lower center is either *Haruna* or *Kongo*. The carrier *Chiyoda* is on the right.

The U.S. fleet, by contrast, suffered only minor damage to one carrier and two battleships—with no ships sunk.

The Japanese fleet began to retreat.

On June 20, Admiral Mitscher took command of the pursuit. He wanted to chase down and finish off the retreating enemy.

But he faced two serious problems:

1. Low fuel—for both ships and planes. The distance to the Japanese fleet was uncertain, so there was a risk of running out of fuel and being stranded at sea.

2. Darkness—it was late in the day. Standard wartime procedure for Allied ships was a strict "lights out" policy at night to avoid detection. But without lights, American pilots returning from late strikes wouldn't be able to find or land on their carriers.

Mitscher made a bold decision. He ordered the fleet to turn on all lights—knowing it would expose the ships to possible attack, but believing it was the only way to bring his pilots safely home.

Just before sunset, Mitscher launched a massive wave of planes.

American bombers located the Japanese carrier fleet and opened fire. One of their targets: the *Zuikaku*, the last surviving Japanese carrier from the Pearl Harbor attack. The strike badly damaged the *Zuikaku*.

Fighter-plane contrails mark the sky over Task Force 58.

The American pilots then turned their attention to the Japanese light carriers *Ryūhō* and *Jun'yō*. Diving from the sky, they dropped their payloads, and one carrier—the *Hiyō*—was hit and sank beneath the waves.

But now the Americans faced a new enemy: the vastness of the Pacific Ocean.

Many of the planes were low on fuel. Some had been hit by enemy fire. With no fuel left to return, pilots were forced to ditch their aircraft into the sea and hope—pray—for rescue. Some were saved. Others were never seen again.

Even pilots who still had enough fuel faced a terrifying challenge: finding their carriers in the dark.

They didn't realize that Admiral Mitscher wanted to give his brave pilots a chance to survive.

The fleet's decks were illuminated, exposing them to potential attack. Mitscher also called for star shells—bright flares—to be fired into the night sky, lighting a path home for the exhausted pilots.

Breaking such procedures could have led to Mitscher being court-martialed—which means put on military trial and potentially removed from command.

But his leadership saved countless American lives. And his judgment proved right. No Japanese forces attacked that night.

The Japanese fleet, commanded by Admiral Ozawa, had been nearly destroyed. It lost:

- 3 aircraft carriers
- More than 400 carrier-based aircraft
- Roughly 200 land-based aircraft
- Hundreds of trained pilots—many of whom could never be replaced.

By the end of the battle, Ozawa, who had started with roughly 500 aircraft, ended with just 35.

The Americans, by contrast, lost 130 planes—many due to fuel running out, not enemy fire. The ships, however, remained almost entirely intact.

The Battle of the Philippine Sea would go down in history as the last great carrier battle of the Pacific War.

More than just a victory, this battle settled the argument that had lingered for decades: which ship was supreme—the battleship or the aircraft carrier?

The answer was clear.

Even Admiral Ozawa had to concede: the aircraft carrier now ruled the seas.

WHO FOUGHT?

Vice Admiral Marc A. Mitscher during World War II.

Marc "Pete" Mitscher didn't grow up on a ship or near the ocean, but in Oklahoma, where his father was once the mayor.

In 1910, after graduating from the U.S. Naval Academy, Marc held a passionate desire for flying. Back then, airplanes were new and dangerous—lots of crashes! Marc became one of the Navy's very first pilots, earning the title "Naval Aviator Number 33."

In 1919, he even tried flying a huge seaplane across the Atlantic Ocean! The plane was forced to land in rough seas before making its destination, but Marc was awarded the Navy Cross for bravery.

Marc focused many years on learning how planes and ships could work together. His knowledge helped design aircraft carriers—the giant ships that became floating airports. He helped figure out how to launch and land planes off the decks, a skill that would change the way wars were fought.

When World War II began, Mitscher was captain of the aircraft carrier USS *Hornet*. In 1942, he led a top-secret mission called the Doolittle Raid.

B-25 bombers took off from Mitscher's ship and flew to Tokyo—something nobody thought was possible—then dropped their bombs. The successful mission gave Americans hope for winning the war, and showed the world that the U.S. Navy could strike back with deadly force.

Later that same year, Mitscher and the *Hornet* also fought in the famous Battle of Midway.

Mitscher's leadership turned the tide of the war in the Pacific. He was later promoted to admiral and given command of the Fast Carrier Task Force—the powerful group of aircraft carriers that led attack after attack in the island-hopping strategy across the Pacific Ocean.

Mitscher became famous for his fast, bold strikes and clever tactics, such as sending fighter planes in first to protect the bombers. And when he ordered his ships' crews to light up the flight decks at night so tired pilots could land safely in the

Philippine Sea, he revealed how deeply he cared about his men and bringing them home.

Though Mitscher was a quiet man and didn't yell much, his men listened and respected him.

After the war, Admiral Mitscher was put in charge of the Atlantic Fleet, but just two years later, in 1947, he passed away. He was just 60 years old.

But his legacy as a fearless flier, brilliant leader, and hero of sky and sea lives on. The Navy has named two ships after him, along with honors in the International Air & Space Hall of Fame in San Diego.

BOOKS

The Philippine Sea 1944: The last great carrier battle by Mark Stille. Fully illustrated with stunning artwork, Mark Stille tells the enthralling story of the last, and largest, carrier battle of the Pacific War, the one that saw the end of the Imperial Japanese Navy as a formed fighting force.

Philippines 1944: Third Fleet's first victory over Japanese land-based airpower by Edward M. Young. This book is illustrated with archive photos, original artwork, maps, and 3D diagrams.

INTERNET

Relive the aerial dogfights and naval clashes of the Battle of the Philippine Sea, thanks to stunning color footage courtesy of the Smithsonian Channel: https://youtu.be/0ovXuSBwktM?si=QewVlvXMyu2yErGv

AFTERWORD

THE AGE OF MIGHTY BATTLESHIPS exchanging broadsides is over.

Today, naval warfare looks very different—and so does the modern Navy. In fact, the U.S. Navy no longer builds battleships at all.

Modern sea battles are now dominated by electronic warfare. Speed, stealth, guided missiles, advanced radar, and satellite technology have replaced the old close-range ship-to-ship combat.

By the late 1990s, the Navy had shifted its focus to more agile vessels like cruisers and destroyers, specially designed for stealth. These ships are built to minimize their visibility to enemy detection, with hidden masts, low-profile radar systems, and sleek, angular designs.

The firepower of today's ships is astonishing. A single modern warship can launch up to 75 precision-guided missiles, striking targets hundreds of miles away with incredible accuracy. Advanced gun systems fire rocket-assisted shells that can hit enemies from over 60 miles out. Radar and satellite systems now allow navies to spot threats from land, sea, and air—long before the enemy is in visual range.

Although nuclear-powered ships still exist, especially in the form of aircraft carriers and submarines, most smaller ships now run on efficient gas-turbine engines, balancing power and speed with cost.

One of the most cutting-edge developments in naval warfare is the Zumwalt-class guided-missile destroyer. With its sharp angles and radar-deflecting shape, it's like the stealth bomber of the sea—nearly invisible to enemy detection.

I hope you've enjoyed exploring some of history's most epic sea battles—from famous clashes that changed the world to little-known skirmishes that still deserve to be remembered.

And I hope this journey inspires you to keep reading and learning about military history, especially the battles fought across the world's oceans.

GLOSSARY OF NAVAL AND NAVAL WARFARE TERMS

Admiral – A high-ranking naval officer, often in charge of a fleet or group of ships.

Amphibious landing – A military assault that begins from the sea and moves onto land, often using boats or landing craft to deliver troops onto beaches.

Armistice – A formal agreement to stop fighting; usually a temporary truce during wartime.

Barrage – A rapid and continuous discharge of gunfire or bombs meant to overwhelm the enemy.

Battleship – A large, heavily armored warship with powerful guns, designed for head-to-head combat with other warships.

Bow – The front end of a ship.

Bridge – The command center of a ship, where the captain and officers control navigation.

Broadside – Firing all the guns on one side of a ship simultaneously.

Carrier (aircraft carrier) – A large warship that serves as a mobile airbase, launching and recovering military aircraft.

Court-martial – A military trial for someone accused of breaking military laws or orders.

Cruiser – A fast, medium-sized warship, lighter than a battleship but with strong guns and armor.

Cutlass – A short, curved sword often used by sailors during hand-to-hand combat in earlier centuries.

Deck – The floor or platform on a ship; what people walk on.

Destroyer – A fast and agile warship designed to protect larger ships from submarines and aircraft.

Escort ship – A smaller warship that protects larger ships, especially merchant ships, from enemy attack.

Fjord – A narrow, deep inlet of the sea between high cliffs or mountains, commonly found in Norway.

Flagship – A ship that carries a commander and bears the commander's flag.

Fleet – A group of ships operating together under the same command.

Flotilla – A small group of naval vessels, usually destroyers or submarines, acting together.

Fuselage – The main body of an aircraft.

Gun turret – A rotating weapons mount on a ship that allows large guns to aim in different directions.

Hangar deck – The area on an aircraft carrier where planes are stored and maintained.

Hull – The main body of a ship that floats in the water.

Keel – The central bottom part of a ship's hull that runs along its length and provides structure.

Knot – A unit of speed used at sea. One knot equals about 1.15 miles per hour.

Landing craft – A boat used to transport troops from a ship to the shore during amphibious assaults.

Mast – A tall, vertical pole on a ship that holds antennas, flags, or sails.

Mini-sub (or midget submarine) – A small submarine used for special missions, often with only a few crew members.

Port – The left side of a ship when facing forward.

Scuttle – To deliberately sink a ship, often to keep it from falling into enemy hands.

Shell – A projectile fired by a naval gun or cannon, usually explosive.

Sonar (Sound Navigation and Ranging) – A technology that uses sound waves to detect underwater objects like submarines.

Starboard – The right side of a ship when facing forward.

Star shell – A bright flare fired into the air at night to illuminate the battlefield or sky.

Strike group – A formation of ships, usually centered around an aircraft carrier, organized for attack.

Submarine (sub) – A warship designed to travel underwater; usually armed with torpedoes or missiles.

Superstructure – The part of the ship built above the main deck (like the bridge, towers, or gun platforms).

Torpedo – An underwater missile launched from a sub, ship, or plane that explodes on contact.

Turret – A rotating armored structure that houses a ship's guns and protects the crew operating them.

Wake – The trail of disturbed water left behind a ship as it moves.

Watertight compartments – Sections inside a ship separated by strong walls to help prevent flooding if part of the hull is breached.

Wolf Pack – A German tactic in World War II where groups of U-boats (submarines) hunted Allied ships together.

"X-Craft" – A British mini-submarine used for stealth attacks on enemy ships during World War II.

Made in the USA
Monee, IL
04 November 2026

67359383R00125